Praise for *Astroqui:*

'A wonderful jaunt through the universe at every scale, and a great way to fill in every gap in knowledge you have about astronomy.' Zach Weinersmith, co-author of the *New York Times* bestselling book, *Soonish*

'*Astroquizzical* is a superb astronomy book, written with a distinctive tone which is both pragmatic and poetic at the same time. It's perfectly attuned to the kind of awestruck curiosity we feel whilst taking in the majesty of a clear, starlit night sky. ... [Jillian Scudder] brings the perfect blend of fact and fascination to help us feel a greater sense of our place within the clockwork of the universe.' Jon Culshaw

'Scudder's mission is to provide the lay reader with a thorough grounding in the basics of astronomical knowledge. ... The writing is fluid and direct with the subject material brought vibrantly to life. ... For astro novices this book ... will bring a welcome depth to their appreciation of the night sky and the wonders it holds.' *BBC Sky at Night* magazine

'[G]enuinely entertaining [with] an excellent balance of enthusiasm and facts ... This is the kind of book that would be excellent to get either a teenage reader or an adult with limited exposure to astronomy interested in the field. It reads well and gives basic details without being patronising. It's a cosmic journey that I enjoyed.' Brian Clegg, *popularscience.co.uk*

'The narrative form that Scudder employs is an imaginary cosmic journey that begins on our home planet and takes us in seven steps to the furthest galaxies. This simple format has been tried countless times before by big-name astronomers. What's different here is an intense level of engagement between writer and reader. Vivid storytelling explains the physics without equations. ... Her aim is to get people to think issues through for themselves, and that works. The clarity of Scudder's writing is impressive.' Simon Mitton, *Times Higher Education*

'[This] excellent debut book is all about making complex concepts, if not exactly easy to understand, then at least a little easier to grasp. ... In her enthralling cosmic journey through space and time, astrophysicist Jillian Scudder discusses our home planet's place in the universe. Beyond the flawless presentation of known facts and current thinking, Scudder explores further by positing counterfactuals and thought experiments. The real triumph ... is that it brings high-altitude, notionally abstract ideas to the general reader, presented in an entertaining and accessible way.' *Engineering & Technology* magazine

'*Astroquizzical* approaches astronomy at a unique angle. It begins by stating that we are all distantly related to the stars; everything we're made of can be traced back to when they explode. By making this comparison at the start of the book, you instantly become intrigued and involved and from then on, the author – Jillian Scudder – does a fine job of covering a variety of topics and interests in space science.' *All About Space*

ASTROQUIZZICAL

A Beginner's Journey
Through the Cosmos

DR JILLIAN SCUDDER

ICON

This edition published in the UK and USA in 2019
by Icon Books Ltd, Omnibus Business Centre,
39–41 North Road, London N7 9DP
email: info@iconbooks.com
www.iconbooks.com

First published in the UK and USA in 2018 by Icon Books Ltd

Sold in the UK, Europe and Asia
by Faber & Faber Ltd, Bloomsbury House,
74–77 Great Russell Street,
London WC1B 3DA or their agents

Distributed in the UK, Europe and Asia
by Grantham Book Services,
Trent Road, Grantham NG31 7XQ

Distributed in the USA
by Publishers Group West,
1700 Fourth Street, Berkeley, CA 94710

Distributed in Canada
by Publishers Group Canada,
76 Stafford Street, Unit 300
Toronto, Ontario M6J 2S1

Distributed in Australia and New Zealand
by Allen & Unwin Pty Ltd,
PO Box 8500, 83 Alexander Street,
Crows Nest, NSW 2065

Distributed in South Africa
by Jonathan Ball, Office B4, The District,
41 Sir Lowry Road, Woodstock 7925

Distributed in India by Penguin Books India,
7th Floor, Infinity Tower – C, DLF Cyber City,
Gurgaon 122002, Haryana

ISBN: 978-178578-412-5

Typeset in Bembo MT by Marie Doherty

Printed and bound in Great Britain by Clays Ltd, Elcograf S.p.A.

CONTENTS

ABOUT THE AUTHOR

Jillian Scudder is an astrophysicist and assistant professor at Oberlin College, Ohio. She has been writing 'Astroquizzical', a blog answering space-related questions from the public, for over five years. Her writing has also been published in *Forbes*, *Quartz*, *Medium*, and *The Conversation*. This is her first book.

astroquizzical.com / @Jillian_Scudder

For the curious

ACKNOWLEDGMENTS

Many thanks to Keaton, for all his encouragement over the years.

Thanks to Mum, Dad, Matthew, and all my friends, who consistently got excited about what I was writing about.

And to all the readers of 'Astroquizzical' over the years, thank you for giving me your curiosity to satiate.

LIST OF ILLUSTRATIONS

Color images in the plate sections

1. The rich star fields of the Large Sagittarius Star Cloud.

2. A pillar of gas and dust located in the Carina Nebula.

3. The eight planets and the new solar system designations.

4. The protoplanetary disk surrounding the young star HL Tauri.

5. A comet-like object called P/2010 A2.

6. A still from the first color movie of Jupiter.

7. Valles Marineris, the largest canyon in the solar system, located on Mars.

8. Hubble Space Telescope photographs of the surface of the dwarf planet Pluto.

9. Image of Pluto from the Long Range Reconnaissance Imager (LORRI).

10. Solar system distances in perspective.

11. Four images of supernova remnants.

12. A Hertzsprung-Russell diagram.

13. Messier 54, the first globular cluster found outside our galaxy.

14. Cassiopeia A, a supernova remnant.

15. The remnant of Supernova 1987A seen in light of different wavelengths.

16. An artist's rendition of a black hole with an orbiting companion star.

17. The large Whirlpool Galaxy and its smaller companion galaxy.

18. Hubble Space Telescope image of the spiral galaxy Messier 101.

LIST OF
THOUGHT EXPERIMENTS

TIMELINE

0 years	The Big Bang: the universe comes into existence[1]
10 seconds–15 minutes	The first atoms form[2]
375,000 years	The Cosmic Microwave Background is produced (the universe becomes transparent to light)[3]
Approx. 180 million years	The first stars in the universe form
Approx. 200 million years	The oldest stars in the Milky Way form[4]
400 million years	The oldest observed galaxy (to date)[5]
3.5 billion years (10.3 billion years ago)	Stars form in galaxies at the highest rate in cosmic history[6]
9.2 billion years (4.57 billion years ago)	The Sun forms[7]
9.3 billion years (4.47 billion years ago)	The Earth forms[8]
13.77 billion years	'Now' relative to the start of the universe

[1] https://map.gsfc.nasa.gov/media/060915/index.html
[2] https://arxiv.org/abs/1706.03138
[3] https://www.nasa.gov/mission_pages/planck/multimedia/pia17449.html
[4] https://arxiv.org/abs/1302.3180
[5] https://arxiv.org/abs/1603.00461
[6] http://adsabs.harvard.edu/cgi-bin/bib_query?1998ApJ...498..106M
[7] https://arxiv.org/abs/astro-ph/0204331
[8] https://pubs.usgs.gov/gip/geotime/age.html

PROLOGUE

You are living, in company with almost every other human, on the surface of the planet Earth, the only planet in the vast universe known to host intelligent life of any kind.

It's often said by astrophysicists that every one of us should feel a strong connection to the stars. Without generations of stars that burned, exploded, or collided before our planet was formed, the carbon that our bodies are made of, the iron in our blood, and the gold and silver of our precious objects would simply not exist.

In a very tangible way, those stars made it possible for us to be here to look at them. Without them, we could not possibly have evolved on our watery world. But truly exploring how we are linked to them – and how they have led to our own lives on planet Earth – can be an arduous task, even for the curious-minded among us. While there are many ties between us and the stars, such information is often forgotten or hard to find.

This book explores the ties that link us not just to the stars, but to the universe as a whole – our cosmic family. Without a planet to call home, we would not exist.

Without a star, our planet would not exist. Without a galaxy, our star would not exist. And without the filamentary nature of structure in the earliest universe, our galaxy would not exist. Each of them paved the way for another generation – building up the groundwork for our tree of life.

Welcome to your cosmic family tree. Let's explore some of the stories that this family has to tell us.

1

A HUMAN PERSPECTIVE

The view from home

As with any hunt for our own human ancestors, we begin our journey up the cosmic family tree with ourselves. How do we see the universe from where we stand? As both the children of this vast cosmos, and the only ones (to our knowledge) who are attempting to chart our cosmic lineage and understand the rules of the universe, our own perspective is a unique one. Our sense of what the 'big picture' is, as well as how we might fit into that picture, is very much affected by how the stars appear to us. As we have developed more advanced ways of viewing the night sky, our sense of just how big the picture truly is has only grown in scope.

Our view of the cosmos is almost always from the surface of our parent planet, Earth. A select few members of our human race have had the privilege of observing our planet from a loftier perch, but with the exception of those astronauts, all of humanity has observed the stars and planets from the ground. It's at night, with our atmosphere protecting us from the freezing void of space, and our own star no longer flooding our planet with light, that we can see the first glimpses of our immediate cosmic family.

Many of us, on a clear night, will look up to find ourselves briefly captivated by the shining of a bright object in the sky. Some of the brighter lights we can see in the night sky are our Earth's planetary siblings, formed out of the exact same cloud of dust and gas that generated our own home 4.5 billion years ago (more on this in Chapter 3). But even without the planets overhead to shine extra brightly, the night sky can be dazzling, especially if you happen to find yourself away from the lights of the city streets.

Without interference from artificial lights, thousands of stars are visible even to our unaided eye, but these skies are increasingly difficult to find. According to a 2016 study by Fabio Falchi, 99% of the US and European populations live under light-polluted skies. It's easy to forget, or to have never seen, just how many stars are visible to us from Earth.

But just as the images from the International Space Station can remind us of the curvature of the Earth, photographers who have the means to travel to the few remaining truly dark places can capture the night sky in those remote spots, reminding the rest of us of what we're missing.

Many photographers aim their lenses at the Orion

Nebula. It's both a very aesthetically pleasing part of the sky, and a very bright one, so it's easy to capture a number of stars in the image. Looking at some of these photographs of the night sky, and then looking at the version above your own homes, it may seem that the images have been exaggerated somehow, or that the number of stars has been digitally increased. This isn't the case – but a camera has an advantage that our own eyes can't access. Our eyes are relatively small light-capturing devices, and we can't increase the exposure time in the same way that you can on a camera to catch even faint light.

Many of these photographs of the night sky (see color plate 1), instead of being exaggerated, come instead from a very long time spent observing the sky with a much larger lens than our eye. The longer you point your camera at a specific part of the night sky, the fainter the starlight you're able to capture. Once the light has been compiled together, we arrive at an astounding view of what our night sky looks like, beyond the limitations of light pollution and the small size of the human eye.

No special filters are required here; many of these photographs are taken by regular digital cameras – slightly fancier than the one in your phone. The astronauts on the International Space Station, for all they have a unique position from which they can take their photographs, have the same technique as the photographers on the ground. If you take a series of 10- to 30-second photos of the sky, you can then assemble your series of images into a single, much more detailed record. The stars that exist in all of the exposures should pop up more brightly in the assembled image, and anything that happens to show up in one 30-second window but not another, will fade.

To capture the faintest stars in an image where many thousands of other stars will appear, you have to take into account the rotation of the Earth. The Earth rotates every 24 hours, of course, and if you want to take images of a single set of stars over the course of several hours, they will be moving dramatically as the Earth rotates us into a different direction. To counteract this, many of the deepest images are taken by attaching the camera to a mount which can pivot as the Earth turns, constantly correcting for the spin of the planet. With this technique, you can take even more images to assemble together, allowing you to bring out the light from fainter and fainter stars as you spend more time taking photos.

The color of the sky

Images of our cosmic relatives also come to us from beyond our planet's surface. Pictures from the Hubble

Space Telescope, for instance, have revealed that the world beyond Earth is a vivid and highly detailed one. But this vividness can be a puzzle; fundamentally, these images are not being taken by a human eye, and in many cases they don't represent what a human eye would see, if we could travel to experience these vistas (see color plate 2).

The human eye has a really unusual sensitivity pattern to light. We're pretty good at seeing things in the yellow-green range, orange we can usually do, but once you get into reds and blues, our eye suddenly gets extremely bad at registering deep reds and dark purples, and our brain translates those colors into 'black', or more accurately, as 'there is no light here that I can deal with'. To anything outside the range of visible light, we are completely blind. This odd sensitivity pattern means that it's quite difficult to make a camera with exactly the same sensitivity as our eye. This is the same reason why it's sometimes hard to get your own camera to pick up the colors you can see by eye. Most cameras have settings nowadays to help change the sensitivity towards a specific color, but they won't perfectly replicate the eyeball's experience.

If you want to make a color picture from an image coming from a space telescope, there is an additional challenge to overcome. All the cameras attached to telescopes are just photon* counters – if a photon makes it through the telescope and into the camera, it adds 1 to the number of photons that arrive from that patch of the sky. This means that the only images you can make are intensity maps – black and white images. For scientific purposes, astronomers are generally more interested in measuring the amount of very specific color slices of light that arrive to the telescope. In order to limit the kind of light that actually makes it to the telescope's camera, filters are usually put in front of it. The filter works in the same way as red–blue 3D glasses and images: the red lens lets through only red light, and the blue lens lets through only the blue, so each eye gets a different picture, and your brain reconstructs the depth of the image.

An astronomical filter is usually constructed to let in light from a very specific physical process – for instance, the color of light that hydrogen produces when it is in a very hot environment. Hydrogen here produces a deep pink color, so instead of a red or blue filter, we'd have a deep pink one. This would let in only light that is produced by that hydrogen, and we can map the locations of that gas on

----

* The photon is the smallest quantity of light – one individual packet (or particle).

the sky. This image is still entirely in black and white, but it is the astronomer's map to untangling what's happening in that part of the sky.

But to reconstruct a colorful image out of this black and white one is no simple task. Given that we're detecting light at much better sensitivities than the human eye, and that we're usually doing it in discrete chunks instead of one (very complex) curve as the eye does, putting these chunks of light back into a single image is a tricky business. Even when all the light is taken from the narrow range that we can see, it must still be reconstructed and tweaked to reflect the brilliance of the colors we've observed. Hubble has produced many beautiful images (such as the nebula in plate 2) labeled as 'visible light images'. What this means is that the narrow ranges of colors that Hubble observed all fall within the range of light detectable by our eyes – but they have still been patched together, the colors of each set of data overlaying each other to build an image in full color. In this particular case it made a lovely and vivid image, but it is still only rendered with six colors, each color coming from its own black and white photograph of the sky. In other words, the images come out of the telescope as black and white, but each is assigned a color and then reassembled. While the general term for this style of image is 'false color', the colors here aren't actually 'false'. The deep pink glow of hydrogen will remain deep pink, and the glow of oxygen, a brilliant aqua, has stayed that color.

'Exaggerated' color images can be used to extend our sight much beyond what we can actually see. Perhaps a galaxy is rather unimpressive in visible light, but has a stunning brilliance in the ultraviolet or X-ray. To our eyes this is dark; but a black and white image from a telescope sensitive to that light can be added to our collection, allowing us to construct an image. In these cases, a color too blue for human eyes is often added as a vivid blue or purple, and a color too red for us is added as a bright red or purple. Sometimes these composites are scientifically instructive, but most of the time they are created to harness the power of an illustration, and are manipulated to reflect the beauty of the image.

Atmospheric problems

It's easy to forget that from the ground, our view of our cosmic family is strongly influenced by the presence of our atmosphere. Even when the skies are clear, our atmosphere can pose some barriers to seeing the stars as clearly as we might like. We humans are not used to thinking of the atmosphere on our parent planet as much of a barrier, in part because we move so easily through it, and it's transparent to the light most of us use to navigate through our world. But simply being transparent to light doesn't mean that light doesn't change as it encounters our atmosphere – and it usually does. If you've ever seen the stars twinkling overhead in the view from home, you've witnessed one of these transformations.

When the stars overhead appear to flicker and change in brightness, that's the atmosphere at work. The stars themselves are quite stable, and aren't changing the amount of light they produce and send towards our little planet. We can check this by waiting a few days – if you go out on a clear, still night, you should find that the stars hang quietly in the night sky, not a twinkle to be spotted. And if you were fortunate enough to observe the stars from space, you would see them as perfectly point-like pricks of light, no matter how twinkly they appeared from the ground.

Back on Earth, if the wind has picked up, you should be able to see the stars twinkling their hearts out. You might see something similar if you look to the stars near a low horizon (sorry, city dwellers). Even if the stars straight overhead seem stable, the ones closer to the horizon may appear unsteady.

Whenever light encounters a gas – and our atmosphere is all gas – the direction the light is pointed changes slightly. How much the light bends depends on a number of things, but the density of the gas is one of them. How densely packed the gas is depends strongly on its temperature, so warm air bends light to a slightly different angle than cool air does. The higher you are in the atmosphere above our planet, the cooler the temperature, but this isn't a completely smooth transition from warm to cold. These temperature changes exist in little bubbles of air, packed against one another.

These little bubbles act like a series of lenses suspended above us, twisting and distorting the light on its way through. If the air is calm, these air-pocket lenses are relatively large, so the light travels through fewer of them, and has fewer deflections on its way down to the ground. Similarly, if the temperature isn't changing rapidly from bubble to bubble, the light won't have to go through as many changes in direction.

This is the same reason that stars overhead might appear to twinkle less than the stars at the horizon. Light from a star directly overhead is taking the shortest path through our atmosphere: straight down. A star close to the horizon is taking one of the longest paths possible, and so the number of air lenses it must travel through in order to reach our eyeballs is much, much higher. With more chances for the starlight to be bent into an unusual place, the likelihood that the starlight will flicker in and out of focus goes dramatically up – these atmospheric focusing problems are what we see as a twinkle in the night skies.

The technical term for this atmospheric interference with starlight is 'seeing'. The better the seeing, the less twinkling the stars are doing. Even if it's a perfectly clear night, you can still have bad seeing, usually due to wind high above us in the atmosphere. If you're after extremely crisp images of objects far away from our parent planet, bad seeing can be a significant problem – with the atmosphere warping pinpricks of light into much larger, wobbly shapes, the images come out of the telescope much more blurry-looking than we would like. If you're trying to distinguish two closely placed stars in the sky, bad seeing can blur them together so much you couldn't tell them apart from each other. This is one of the main reasons that astronomers like space telescopes, even with the difficulty of placing them there. There's no blurring of the distant starlight in space.

Beyond simply blurring light, which is a rather intangible influence of our atmosphere, the air which surrounds our parent planet also serves as a very effective barrier to small physical objects coming our way, and there's no better example of this than a meteor shower.

Meteors

Meteors are little pieces of stuff – usually pebble-sized pieces of rock – that have the misfortune of running into our planet. The name meteor distinguishes them from objects in space, which are meteoroids, and bits of rock that actually survive the passage through the atmosphere and reach the surface, which are called meteorites.

Running into the atmosphere of our planet spells doom for most small objects. The change from the void of space to the relatively high density of the gases of our atmosphere means that these pieces of the solar system are rapidly slowed down, like an arrow burrowing into a straw target. As they slow, they donate energy to the gas surrounding them, which heats up and vaporizes the outer layers of the rock as it plunges groundwards. If the meteor is small, this process evaporates the entire meteor in the blink of an eye, and the flash of its glow fades from sight. This can happen at any time of day, but we associate them more with the night skies, because their luminous ends are much easier for us to spot when the light from the Sun isn't there to compete. However, if the meteor is big enough, it won't matter whether the Sun is up; the fireball that exploded over Chelyabinsk, Russia in 2013 was perfectly visible in the morning Sun.

There are particular times of year when the odds of

spotting a meteor are much higher than normal – these nights are what we call a meteor shower. The odds go up because the Earth is passing through a particularly pebble-filled patch of our orbit, usually the remains of a comet which passed by many years earlier. As the comets, which are made of rock and ice, come close to the Sun, the ice melts, freeing the rock into space. These small rocks and other pieces of debris remain behind the comet's path, the way a particularly muddy dog is easily traced through a house. When the Earth catches up with this path, we're in for a very small atmospheric pummeling, and a fun light show in the night sky.

The aurora

A meteor shower is certainly a dramatic nighttime event in the skies, but it is outdone by the aurora – the Northern or Southern Lights. Our planet doesn't have a monopoly on the aurora – other planets with magnetic fields surrounding them can also produce them. But even though we're one of a family of planets with the aurora, ours certainly puts on a delightful spectacle.

There are only a few things you need to produce the aurora: dark skies, a magnetic field, and an active star. Our home planet comes equipped with a magnetic field, generated by the motions of the metals in the Earth's core, so we've taken care of that one right away, and dark skies come every twelve hours or so. An active star is also a regular occurrence, and what we need from our star, the Sun, is a large volume of charged particles (an electron or proton will do – see pages 108 and 127).

If the Sun provides these high-energy particles so that they hit the Earth, our magnetic field deflects the majority of them away from the planet before they ever come in contact with the atmosphere. (This is good, because without a magnetic field, this is an effective way of losing an atmosphere.) However, if there are particles that come towards our planet aimed at the poles, the magnetic field is less able to keep them away from the atmosphere. Near the magnetic poles, the magnetic lines that normally run parallel to the surface turn and sink into the surface. This creates a divot in the magnetic shield, and particles can get stuck in there, like bits of leaves in an eddy. The solar particles end up smashing into the gas of our atmosphere. The energy donated to the atoms★ of gas makes a glow, lighting up the skies with the aurora.

....

★ An atom is the smallest uncharged particle in the universe, and combinations of atoms make up all the materials on our planet and in the rest of the universe.

If you're particularly keen to see the aurora, there are a few things you can keep an eye out for. As with all faint astronomical displays, it is easier to spot the aurora at night, and the darker your skies, the more visible they will be. The other factor is how close you are to the magnetic poles. The closer you are, the less of a donation of particles from the Sun is needed to see the lights. With very strong storms of particles, the aurora can be visible further from the poles, but these storms are few and far between.

If it's starting to sound to you like all the good astronomical vistas from our parent planet are visible only when our illustrious Sun isn't around, I can't blame you, but there are certainly a few things that stick around during the day. The Moon is easily visible in the daytime sky, and if you know where to look, Venus is also bright enough to be seen while the Sun is up (various apps for your phone can point you in the right direction). There's nothing special about the Moon and Venus here, they're simply reflective enough to be visible – in principle, any kind of object in our skies could be seen from the ground during the day, assuming it can shine enough light down to the surface.

THOUGHT EXPERIMENT:
Sights by day

To put reflectivity in a more tangible context, let's take one of the largest crafts we use to cruise the seas of our planet, and put it in space, to cruise a very different ocean. It's made of metal, and fairly sizable, so if we put it in orbit around the Earth, it should have a good chance of being reflective enough to be visible from the ground.

A supercarrier, which is among the largest aircraft carriers, typically measures 77 meters wide by 333 meters long. An object of that size has an area of 25,641 square meters, or 2.56 square kilometers. In order to know how much light it would reflect (and thus, how bright it would appear), we have to know what it would be made of. On Earth, a lot of our marine craft are made of steel, but spacecraft don't need to follow ocean rules exactly,* so we could consider making our supercarrier into an aluminum supercarrier. Aluminum is one of the most reflective metals out there, reflecting 91% of the light that hits it. It's also quite light, which is nice if you need to get it into space from the ground.

We do have one point of comparison to work with – the International Space Station (ISS) is already in orbit, and sits at 72 meters wide and 108 meters across, counting all of its solar panels. The ISS is therefore about

---

* In fact, they probably shouldn't. Other than being waterproof and airtight, which they *definitely* should still be.

7,800 square meters, or about 0.78 square kilometers. So a supercarrier would be 3.3 times larger in reflecting area than our current largest spacefaring residence. The ISS reflects around 90% of the light that hits it – this is on purpose, to help keep the space station cool, and at less than a square kilometer in size, and at 90% reflectance, we can see the ISS with the naked eye, *if* it is at its closest to the surface of the Earth.

Since we're dealing with a much larger surface area, that's a good hint that a supercarrier-sized object would also be visible to a watching public on the ground, and that's true, as long as the material our spacecraft is made of is similarly reflective. With the larger surface area, we'd expect an aluminum space-supercarrier to reflect 3.36 times as much light as the ISS does.

In terms of spotting such a craft, this kind of object would be easily visible at night – it's between 4.8 times (for steel) and 7.7 times (for aluminum) as bright as Venus. (If you can see *any* starlike object in the twilight or early morning sky, it's Venus.) A spacecraft of this size would appear as an extremely bright star moving relatively quickly through the night sky. During the daytime, where the Sun dominates our sky, we'd still be able to see it. At more than five times brighter than Venus, this sort of spacecraft would certainly be visible from the ground. Any kind of spacefaring future for our species will likely fill the skies of our planet with plenty more bright objects like this.

Human visibility on planet Earth

The inverse is also true – while floating in space, it's relatively easy to spot human activity on our planet. The easiest way, of course, is to look for the lights of our cities at night. The astronauts on board the International Space Station have taken some beautiful images of this, with the grid pattern of the roads in the US contrasted with the more tangled street patterns of older cities in other parts of the world.

If we were to go hunting for life on another planet, modifications of that planet's surface would be a big thing to look for, if we had the resolution to go with. (This would be an 'orbiting the planet' hunt, and not a 'searching from here on Earth' hunt.) On Earth, we humans have not only illuminated the night, but we've carved into the fields, adjusted coastlines, and irrigated dry terrain.

But science fiction likes to take a different approach. Instead of searching for cities by their lighting, there's a vaguely phrased 'signs of life' method, which often translates to hunting for signatures of warm bodies on the planet. If we turn this lens onto ourselves, we're certainly visible through a heat-based search, but only if you're close enough to the planet.

How many of the human-induced changes to our planet's surface you'd be able to see from space depends entirely on the resolution you can achieve with your camera – how small an object can you spot? Resolution for an image depends on only three things: how close you are to the object in question, what wavelength of light you are looking at, and how many wavelengths of that light you can fit across your telescope. If you're looking for warm bodies, you'd need a heat map. That means you're looking in the infrared, from at least an orbital distance around the planet. How much can you see in the infrared?

In general, as you would expect, in infrared the poles of our parent planet show up as cold, and the equatorial regions as much warmer, but at this resolution, you can't see any real details. Cities don't show up here, let alone individual humans. This is due to the combination of the wavelength (the infrared is a longer wavelength than optical light, so the resolution drops), the distance the satellite is orbiting the planet (about 440 miles up), and the size of the collecting area of the satellite.

You *can* spot cities via infrared heat measurements; if you're not in the desert, dense cities tend to be warmer than the surrounding areas. Part of this is that we've cut down all the trees to build the city; another reason is that we've paved it with heat-absorbing asphalt. If the city has a lot of trees planted, this city 'heat island' is less obvious. The resolution on these heat-island images is about 100 feet, which is still well too large to detect individual people. The resolution here is partially because the size of the mirror on this satellite is still only 16 inches across (not very big, in the scheme of things).

If you just want high resolution to capture the smallest details on the surface of our planet, the best bet is to bring a really large mirror and camera (increased collecting area = better resolution), or to swap over from infrared to the optical range, though clouds will become a problem if you do the second one. On Earth our cloud layer is not very thick, not very hot, and tends to move over time, so if you wait long enough you should be able to see what's underneath any given cloud sooner or later, but if you're observing a planet more Venus-like in its permanent cloud cover, the optical is not going to be your friend.

On Earth, however, it works fine; commercial satellites in orbit can now image the Earth down to a resolution of about a foot. (Or at least that's as good as various militaries will allow them to disclose; super high-resolution imagery of the Earth's surface is also used for military reconnaissance.) With optical high-resolution data, you can look for geometric patterns. Perfect circles, squares, rectangles, or triangles are unlikely to happen naturally, so if you spot widespread rectangles on the surface of the Earth, that usually means you've found a well-planned city or a farm, either of which indicates some kind of intelligence at work.

Of course, the further away from the planet you are, the harder this is to do – it's not the sort of scanning you can do while cruising the galaxy at high speeds. To map the whole Earth at low resolution (between around 800-foot and 3,200-foot resolution, or 250–1,000 meters), the MODIS instrument on one of our Earth-orbiting satellites, orbiting at about 450 miles above the surface, takes two days. So it's possible to detect signs of life on a planet via heat-based images if we're looking for evidence of cities, but not if we're looking for individuals, and not if you don't want to spend a few days in orbit around the planet.

Still, if you built a *very* large-aperture, wide-angle telescope, and had it orbiting the planet in space, you just might be able to spot people outside. If you had a telescope that was 500 meters in diameter, you'd get resolution that's 150 times better than what the Hubble Space Telescope can do. Even in the infrared, we'd be able to detect individual human beings if we gathered that much light – though to tell that anything was moving, you'd have to take a series of images and play spot the difference. (A series of extremely short exposures would also keep all your images from being blurred into unrecognizability, unless you've parked the telescope in geostationary orbit.) If you had an inkling of where to point your dish – and weren't reliant on mapping the entire planet – the civilization we've dreamt that Earth becomes in the future may yet be able to spot intelligent life walking around on other planets.

THOUGHT EXPERIMENT:
No childhood pictures of the Earth?

Unlike a dig through your own family history, if we go looking for old pictures of the Earth from ground-level, we find ourselves very rapidly limited to the window of time after cameras were invented. To push earlier, we look to illustrations of life on Earth, which brings us shockingly early into human history; humans have been drawing their surroundings for an astoundingly long time.

If, however, we're interested in photos of the Earth as a whole, then our window of opportunity declines even further, because we are limited to the slice of time between the present and our first Earth-monitoring satellites. The oldest photos of our planet from space are from 1946.* By geological or astronomical standards, we've taken a rapid-fire burst of images, but we will never unearth a photo of our planet as it was 200 years ago, the way you might discover a photo of your human great-grandmother in the 1850s.

* https://www.nasa.gov/multimedia/imagegallery/image_feature _1298.html

It's an appealing idea to be able to see the Earth as it was, many years past. With a curious mind, we could wonder – is there *any* way that we could see the Earth in the past? Is there a way we could harness the delay that comes with light's travel through vast distances to see the Earth hundreds or millions of years in the past? Geometrically, this is *possible* – reflected sunlight from our Earth could go out into space, be reflected off a gigantic mirror, and back towards us.

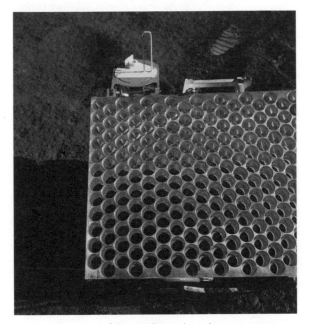

1. A portion of the Apollo 15 lunar laser ranging retroreflector array, as placed on the Moon and photographed by D. Scott. (NASA/D. Scott)

We actually do use this bouncing time delay already (at a much smaller distance scale), because a few of the Apollo missions put mirror-like reflectors on the surface of the Moon (see image above). They're not the standard type of mirror, but a special reflective surface called a retroreflector, which directs light back the way it came from, which normal mirrors don't do. By firing high-powered lasers towards the surface of the Moon where the mirrors are situated, we can count the time delay for the round trip of the laser beam. This is primarily used to measure the distance between the Earth and the Moon to incredibly high precision. These measurements are one of the methods by which we know that the Moon is receding from us by a little more than an inch every year.

For a laser, this round trip from the Earth to the Moon takes a relatively quick 2.5 seconds or so to go the 478,000 miles (roughly) there and back. Even though we're not dealing with large distances (on an astronomical scale) yet, there's a hint of the problem we might face with our hunt for ancient photographs of our parent planet. We start to run out of photons.

Even a laser, which starts out with all of its light focused into a very small beam, will spread out at larger and larger distances. Red lasers are more prone to doing this than green lasers, simply because the wavelength of light is shorter for green lasers, and this spreading is partially a function of wavelength. (Purple lasers would be even less prone to spreading out.) However, over several hundred thousand miles, even the highest wavelength lasers we can manufacture are going to spread out, and by the time this light gets to the Moon, the laser is only able to faintly illuminate the surface, and only a tiny fraction of that light is going to be bounced off the reflector and back to Earth. The further away you put your mirror, the worse this problem gets, because light gets more and more spread out the further it's traveled.

The Earth does reflect sunlight out into space, so we're not in trouble there. However, this light begins to spread just as the laser light does, but since it's coming from a larger area to start with, it's never as compact and focused as the laser light. By the time that we get to any distance away from the Earth, this reflected 'Earthshine' is very dim indeed. We can see Earthshine if the Moon is up at night. The reason the dark part of the Moon isn't 100% black is that it's getting some reflected light from the Earth. By the time this Earthshine light travels for several hundred light years, one can well imagine that it has become very, very diffuse. And then, of course, it would have to travel several hundred light years back, becoming even more diffuse on the return journey. We'd also have to conveniently find a mirror out in space which has a clear light path between us and it – and if we just put it out there now, we'd have to wait a few hundred years for anything to come back our way.

But let's say we managed to get a few photons back from our several hundred light years-distant reflector, which we can arbitrarily make sufficiently enormous that this would happen – would we be able to identify them? Part of the reason we like using lasers for our Moon experiment is because they're all of a very particular color of light, so we can count up the returning photons at that color, relative to photons of any other color, which we know are unrelated to our experiment. The Earth is *not* a single color, and the atmosphere is incredibly complicated, so the set of photons that we would reflect would be a much more complex set than the laser beam we're firing at the Moon.

On top of this, when the Earth is showing the most reflected light, it's because the angle between us and the Sun is the smallest. So when the Earthshine is the brightest, we're also most likely to be blinding our reflector with light from the Sun. The Sun is really, *really* bright. Stars in general tend to be a big problem for taking direct pictures of planets around other stars, because they're so bright that they swamp out any of the reflected light from a planet, and we have to get really clever with how we block out the light from the star without blocking out anything else.

So, to make sense of our parent planet's past we make do instead with the stories the Earth can tell us based on other measurements — without photographs, we have to rely on geology, planetary science, and theoretical models.

✳

Even though the majority of our human history has seen us observing the cosmos from a grounded place, we nevertheless have had a perpetual reminder of our planetary family, in the form of our constant companion, the Moon. Our Moon, suspended as it is in our skies, is Earth's closest family member, and the trigger for much of our human curiosity about the cosmos. In the next chapter we'll move away from the Earth and look in more detail at our cosmic companion.

2

THE MOON

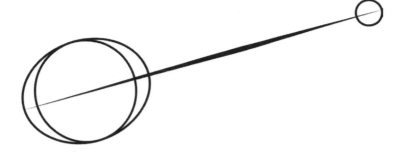

The Earth's cosmic companion

The Moon is simultaneously a constant and ever-changing presence in our skies. As the Moon moves in its orbit around our planet, and the angle between it and our Sun changes, the illuminated portion of the Moon is perpetually creeping along its surface. And yet, if you know where and when to look, it is always a feature of both our daytime and nighttime skies.

The Moon was formed, according to current theory, when a large object, about the size of Mars, slammed into the early Earth a few tens of millions of years after the solar system began (still about 4.5 billion years ago), shredding its outer layers, and destroying the other object. The debris from this astoundingly violent encounter would have formed at least a ring, and possibly a haze of very hot debris, around our planet. As that debris cooled, it collected back into the object we now know as the Moon. The Moon is built out of the same rock as much of the

Earth, as a result of this collision, and it is a much larger moon than you might expect a planet the size of the Earth to have. The Moon has been our companion since that early collision, and our nearest and most visible reminder of the expanse of space beyond our planet throughout all of human history.

The constancy of the Moon's companionship is reflected in another way, to our eyes on the surface of the Earth. We always see the exact same side of the Moon. It's always the same, familiar features of the lunar landscape which face our little planet, no matter what time of day, what the phase of the Moon is, or whether there is a solar eclipse. The features we see at the full Moon, where the Earth-facing side of the Moon is fully illuminated, are the only ones we'll ever see from here. A half Moon appears when only half of the near side of the Moon is illuminated, with the other half plunged into shadow. The unobservable far side of the Moon is also half illuminated, half shadowed.

The far side of the Moon is an almost impossibly foreign-looking place. It's entirely full of craters – unlike the near side, where the dark patches of the Moon (known as 'mare') look relatively clear of craters, the far side of the Moon has no uncratered space left.

The reason we never manage to see this 'other side' of the Moon is because of a trick of the Moon's orbit around

2. This outstanding view of the full Moon was photographed from the Apollo 11 spacecraft during its trans-Earth journey homeward in July 1969. When this picture was taken, the spacecraft was already 10,000 nautical miles away. On board Apollo 11 were commander Neil Armstrong, command module pilot Michael Collins and lunar module pilot Buzz Aldrin. While astronauts Armstrong and Aldrin descended in the lunar module Eagle to explore the Moon, Collins remained on the command and service module Columbia in lunar orbit. (NASA)

our planet. The Moon has settled into a rotation and orbit that match each other – for every trip around the Earth, the Moon rotates once. One full 'day' of the Moon takes about four weeks – the same length of time it takes to complete a cycle of the phases of the Moon. If the Moon didn't rotate, we would see all sides of it, including its strange cratered far side. The only way for us to see such a constant face of the Moon is if it's also very slowly rotating.

Think about sitting in a spinning chair, and holding a book at arm's length. If you hold the book, and spin yourself around, you will always be able to read the text within the book. But the book's orientation with respect to the *room* has done a full circle as you spun. If you asked a friend to hold the book, and walk around you but always face the north wall, after half a spin, you'd be looking at the spine of the book, not the inside text.

The configuration of rotating to match an orbit, the way the Moon does, has been dubbed 'tidal locking', and is simply defined as a setup where one face of an object is continually facing another as it orbits. Sometimes both objects can be tidally locked to each other, which isn't the case for the Earth. If the Earth were tidally locked to the Moon, the Moon would never rise or set in our skies, and half the planet would never see it at all. But this is the situation for the dwarf planet Pluto and its largest moon Charon – both are tidally locked to each other. The amount of time it takes to orbit around the planet will vary from object to object (Phobos, one of the moons of Mars, is tidally locked and orbits Mars every eight hours – way faster than our Moon), but as long as the object is tidally locked, the length of the 'day' on that moon will match the length of time it takes to go around the planet it orbits.

Tidal locking is a remarkably stable setup. Now that the Moon has got into that configuration, it will remain there for a long, long time. The reason why it's so stable is buried in the physics of tidal forces.

Tides

For those who live by the sea, you may already know that the Moon is what creates the ocean tides. High and low tides are the ocean's response to the gravitational pull of our Moon. A glance to the sky can therefore tell you if the tide is high or low. If the Moon is overhead, you're at high tide; if it's at the horizons, your shores are at low tide.

We have ocean tides because the Moon is massive, and relatively close by, and so the gravitational force from the Moon on the Earth is significant. In astronomical terms, if the strength of a gravitational pull changes dramatically across a specific object, that's a tidal force. The Moon's gravitational pull is the prototype for this. The pull of the Moon on the near side of the Earth is significantly stronger than its pull on the far side, because Earth and Moon are so close together. There are no limitations on what kinds of objects can experience a tidal force. A human could feel a tidal force if you went too close to a black hole. A comet can feel a tidal force if it goes too close to Jupiter, and a moon can feel a tidal force because of its planet.

On Earth, the presence of the Moon means that the side of the planet facing the Moon at that moment is feeling a particularly strong gravitational pull, and the exact opposite side of the Earth feels a particularly weak pull. If the Earth were softer, this would stretch it out into more of an elongated shape, towards the Moon. However, the Earth is made of rock, and this material isn't easily stretched. The rocky material just has to deal with a certain amount of strain. On the other hand, the water coating our planet is extremely easily shifted around, and it is free to rearrange itself in response to a stronger pull on one side of the planet, or a relatively weaker pull at the opposite side. So this stronger pull on the Moon-facing side results in a high tide underneath the Moon. But paradoxically, high tide also exists at the weakest gravitational pull – the exact opposite side of the Earth. This weakest gravitational pull allows the water to drift away from the Moon more than it would be able to on the Moon-facing side, and so the water there can hang back, relative to the rocky sphere of the Earth, which pretty much stays put, thus forming another high tide. Low tide, meanwhile, occurs where the oceans are being stretched *from*, exactly in between the two high tides.

The air in our planet is also relatively free to move around – as a gas, it doesn't have the same pressure to remain fixed in place like the rocky underbelly of our planet. Our atmosphere also has regular swellings and sub-sidings, much like the ocean tides, but it's the Sun that triggers these changes, instead of the Moon. While we also call these changes to our atmosphere 'tides', it's in the older sense of the word (a regular change), and it's not the Sun's *gravity* that's making such a large impact.

You can guess that the Sun's gravitational impact on atmospheric tides must be small by looking at the ocean tides. If the Sun were a major player in the ocean tides, then high tide would happen at local noon every day, instead of varying based on the Moon's position in the sky. The Sun is still a player – the highest tides occur when the Sun and the Moon align so both of them are pulling in the same direction – but the difference between a normal high tide and high tide when all forces align isn't as large as the difference between high tide and low tide; the Moon wins that round.

The key difference between atmospheric and ocean tides is that, unlike our liquid oceans, the atmospheric gases have some extra things that they rely on. Liquids are fairly straightforward under normal conditions; they fill their vessels, and are relatively constant in density – it's hard to compress or expand a liquid by very much. Gases, on the other hand, are highly variable in density. It's fairly easy to compress a gas – we do this simply by talking. Gases are also incredibly sensitive to temperature; one of the easiest ways to make a gas more dense is to simply cool it down. Conversely, if you want to make a gas puff up and take up more space, heat is an easy way. Each particle of gas gains a little more energy than it had before, and it bounces around a little faster, and so the whole thing winds up slightly more puffed up than it used to be. So the Sun, as an enormous source of heat, has a pretty major influence on the gas in our atmosphere by simply heating it up.

When the Sun is shining on the atmosphere (in other words, anytime it's daytime), the Sun heats up the gas that surrounds our planet and makes the whole volume expand outwards towards space. This expansion can be measured at sea level as a very slight reduction in atmospheric pressure, but is a lot more dramatic at higher altitudes, where the density of gas is really low to start with – as the atmosphere underneath it expands, suddenly there's an upwelling of denser gas from below. This particular effect isn't a gravitational tidal force at work, but the official terminology is an 'atmospheric tide', and it is a regular, cyclical change.

Of course, the Moon *does* have a gravitational tidal effect upon the atmosphere, but like the Sun's impact on our ocean tides, it's a much weaker effect than the heating provided by the Sun. If the Moon were the main cause behind this atmospheric stretching, it would work the same way as the ocean tides. High tide would mean that you also had the most atmosphere above you, instead of what we see: a 24-hour cycle of our atmosphere heating and cooling under the Sun's rays.

Let's go back to the idea of tidal locking, now that we've got a better handle on what tides can do. These same gravitational tidal forces that distort the oceans do more than just that; they can also resist the rotation of an object. When the Moon was very young, it would have rotated at a much faster speed, and probably would have orbited the Earth at a different speed. The Moon was almost certainly not tidally locked when it first formed. Had any observer been on that early Earth, they could have seen all sides of the Moon as it spun.

The gravitational pull from the *Earth* – which, like the tides due to the Moon, pulls on the side of the Moon closest to the Earth more than the far side – resisted this faster rotation. This resistance due to the gravitational pull of the Earth gradually slowed down the faster spin of the Moon until it was no longer rotating faster than it was orbiting. Once the Moon's rotation had slowed so much that a single face was always facing the surface of the Earth, that was the beginning of its tidally locked orbit, and the Moon has stayed in this configuration ever since.

The Moon also has the same influence on the Earth, but since the Moon is so much less massive than the Earth, this resistance to rotation takes a much longer time to impact on the Earth's spin. However, it's still a measurable effect! The Moon is slowing down the rotation of the Earth by about 15 microseconds every year, gradually lengthening our days.

Supermoon and Micromoon

Not only is the Moon tidally locked to our parent planet, but its orbit happens to be *almost* exactly circular, which means its distance from us doesn't vary that much over the months. Because the Moon isn't going on long, looping journeys away from the Earth, our view of the Moon is of an object that is very close to a constant size on the sky.

Without something catastrophic happening to the Moon, like a giant impact, we wouldn't expect its physical size to change. Because it is made of rock, like the Earth, it won't evaporate over time, even in the harsh environments of outer space. Radiation from the Sun does bleach all the surface material on the Moon, but it shouldn't remove any significant amount of material. If something catastrophic *did* happen to the Moon, it would certainly fling material away from its surface. However, large impacts are increasingly rare in our solar system – we effectively ran out of all the chaotic large objects in impacts much earlier in the lifetime of the solar system. This means that a very large-scale event is extremely unlikely (although not impossible).

There can be minor changes to the Moon's size on the sky, though. The Moon is on a slightly elliptical orbit, which means it has a closest and furthest distance from the Earth as it goes around us. The difference between the closest and furthest approach is about 42,000 km. Considering that, on average, the Moon sits a solid 384,402 kilometers away, this is a relatively minor shift in distance. When the

3. A lunar distance, 384,402 km, is the Moon's average distance to Earth. The actual distance varies over the course of its orbit. This image compares the Moon's apparent size when it is closest to and farthest from Earth. (Wikimedia user Tomruen; CC BY-SA 3.0)

Moon is closest to us in its orbit, and also full, that's when you get a 'Supermoon', but as you can see from the picture above, it's not that big a change from the furthest and smallest the Moon ever gets from us (the 'Micromoon') – the size difference is about 12%.

On top of this orbital change, because the Moon is broadly interacting with the Earth gravitationally, the Moon and the Earth are slowly exchanging energy. This interaction is responsible for the slowing of the Earth's rotation, as we saw, and additionally allows the Moon to drift away from the Earth. This happens at an incredibly slow rate: we're talking about 40 millimeters (1.5 inches) every year. Compared to the normal orbital variation of 42,000 km, this effect is barely noticeable, and certainly won't be to the human eye.

But, the next time you hear something in the news about a Supermoon, remember that it's the closest the Moon will ever be to the Earth, if only by a few millimeters. By the same token, each Micromoon gets a few millimeters more distant, and a tiny bit smaller, each year.

Small-scale pummeling

While large objects are relatively rare in the inner solar system, small objects certainly abound. Without a protective atmosphere like the Earth's which serves to burn up the smallest pieces of grit around our planet, the Moon does not get to enjoy meteor showers. Any meteoroids or other objects on a collision course with the Moon can go crashing straight into its surface, with no slowing whatsoever.* Inside of our protective parental shell, it's easy to forget how harsh the solar system can be, and one of the reasons it's so harsh is a constant flow of cosmic rays, usually coming into our solar system from elsewhere in the galaxy.

Cosmic rays are actually highly accelerated tiny pieces of grit and not rays of light at all. Many of them are subatomic particles in the form of protons, constantly streaming throughout the universe. Whenever they smash into the surface of the Moon, they release so much energy that they can produce an extremely high-energy form of light: gamma rays.

Gamma rays are one of the highest-energy forms of light in the universe – their wavelength is around a picometer (that's one trillionth of a meter, or 0.000000000001 meters), which is roughly the same order of magnitude as the size of a hydrogen atom. These wavelengths are a thousand

* There's a whimsical term for this kind of impact. 'Lithobraking': to slow down by hitting rock. Usually this ends badly for the fast object.

times shorter than visible light, and therefore invisible to our eyes.

The high energy of gamma rays means that this form of light is extremely damaging to life. Ultraviolet (UV) light, which is also too high-energy for our eyes to absorb, though considerably less energetic than gamma radiation, is already damaging to the human body. UV light doesn't penetrate far into your skin – it can damage the surface layers, giving you a sunburn (and increasing your risk of skin cancer), but it won't damage anything past the first few layers of skin. Gamma rays can penetrate much further into your body, and can destroy or alter the DNA within your cells, causing drastic changes to the replication instructions of those cells. This can cause radiation sickness and/or cancer to appear, but without any surface burning of your skin. Very fortunately, our Earth's atmosphere is opaque to gamma rays. Our atmosphere does a wonderful job of protecting us from the beating our cells would otherwise be taking.

On the other hand, the fact that our atmosphere is such an effective wall to this wavelength of light means that we can't observe any of the gamma rays produced elsewhere in the universe, or on the Moon, from the surface of the Earth. In order to look at this light, we have to use a space telescope which avoids the interference of the atmosphere.

At the moment, the Fermi Gamma Ray Space Telescope is observing the whole sky for gamma ray sources, which can tell us about our own galaxy, other galaxies, or the explosions of massive stars in supernovae.

Because the Moon is *constantly* being bombarded by cosmic rays, this means that it is also continually producing gamma rays as the cosmic rays come to a screeching halt. To a space-based observatory, this means that the Moon glows in gamma rays. The Moon, as viewed by a gamma ray telescope, is actually brighter than the Sun – which, very fortunately, does not produce much in the way of gamma radiation. If it did, the gamma radiation from such a close, bright source might have destroyed our atmosphere, removing our shelter from this cosmic hazard.

This kind of cosmic pummeling is something we will have to guard against for any kind of future travels outside our home world, and one of many reasons that underground structures on other planets have been suggested as better places to put our astronauts than the surface – that way the ground can take the hits instead.

An echoless Moon
The complete lack of an atmosphere to protect the Moon has some other interesting implications – on top of being a

world without meteor showers and aurora, it's also a silent world. For all the chasms, craters, and horizons that future human explorers might get to discover, no sounds will await them on the Moon. This means, among other things, that nothing a lunar explorer could do would replicate the sensation of an echo off a mountain range.

Echoes happen on Earth when a sound wave bounces off of something and reflects back towards you. This is most

4. An oblique view of the crater Daedalus on the lunar far side as seen from the Apollo 11 spacecraft in lunar orbit. This is a typical scene showing the rugged terrain on the far side of the Moon. (NASA)

noticeable in canyons, or other places where there's a big, relatively flat wall at some distance away from you. If you shout, you'll hear a faint response, and the delay depends on how far away you are from the wall that's reflecting the sound.

But echoes are all caused by sound waves, and sound always needs a medium to travel through. Fundamentally, sound is a compression wave; if there's nothing to compress, there's no way for the wave to move. Even stomping around wouldn't generate that immediate aural feedback – the sound of stomping would simply not exist, except within the space suit of the explorer. That's not to say that stomping wouldn't vibrate the things around you – they certainly would; it's just that you wouldn't be able to *hear* them vibrate the way you can on Earth.

THOUGHT EXPERIMENT:
Interplanetary transportation?

The Earth and the Moon are tremendously different environments, and perhaps nothing illustrates the differences between our parent planet and its companion more than the thought experiment of opening a window between the two worlds, and allowing humans to step between them. Or rather, we can imagine what kind of chaos would ensue if we could have an open wormhole between the two.

There are a few problems. Firstly, as we've mentioned, the Moon has no air. If you place a human in this wormhole window, and give them something to hold onto, you're not likely to have problems with the lack of air on the Moon, because the Earth's air will be flooding through the window, to the vacuum of space. The daytime temperature on the Moon (224°F/106°C) would be a problem for a human under normal circumstances, but if you're surrounded by the vast rushing of air flooding outwards to the Moon, the Earth's air will again buffer you against this sort of thing. (If you open your window onto the Moon's night, you might end up getting snow as the air hits the chilled lunar surface.) All the wind chill equations I could find indicate that room temperature air (starting at 70°F) will only get down to 60°F or so, no matter how fast it's rushing out of a window. With an abundance of air in our window-view of the Moon, we should be okay to breathe, but there are other problems coming our way shortly.

The main problem is the wind speed. There's a handy equation called the Ensewiler formula to convert a pressure difference into a wind speed, which is written out as $P = 0.002496 \, v^2$. P is the pressure difference in pounds per square foot, and the velocity comes out in miles per hour. An opening between the Earth and the Moon is a window between standard air pressure (1 atmosphere, or 2,116.216 pounds per square foot) and no atmospheric pressure at all. So if we put a difference of one atmosphere into the Ensewiler equation, we get a velocity of 920 miles per hour.

920 mph is kind of a problem. This is 1,480 km/h, or 411 m/s, and faster than the speed of sound. For some context, the fastest recorded wind speeds on Earth are 253 mph (recorded in Tropical Cyclone Olivia, which hit Australia in 1996) and an F5 tornado that hit Oklahoma in 1999, which clocked in at 302 mph. Felix Baumgartner's jump from the edge of space in 2012 got him up to a speed of 843 miles per hour before he pulled his parachute, but he was wearing a massive protective suit, specifically designed to keep him safe, and he crossed the sound barrier quite high in the atmosphere, where the air was not very dense. Unprotected, the shock from entering a 920 mph wind won't kill a human immediately – a shock wave from a bomb only becomes lethal at about 1,500 mph. It's still not doing anyone any great favors, but you won't die from the pressure impact.

The biggest challenge would lie in how to restrain yourself on the surface of the Earth, without being flung headlong into space by the sheer wind pressure. If you tried to keep yourself in place by stapling heavy-duty handles to the Earth-facing side of the window and holding yourself against the wind, you'd rather rapidly run into some problems. The average person has grip strength of approximately 500 Newtons of force. The drag force from 920 mph against a human body is 15,154 Newtons in the other direction. 500 Newtons is not going to be enough to keep ahold of anything against that pressure. 500 Newtons will keep you holding on to something in the face of 182 mph (293 km/h) winds, but nothing more than that. No matter how much a person clung on, the force of the wind would blow them away.

This calculation assumes that our human daredevil is able to continuously grip the handle until the wind speed overcomes their grip strength. In actuality, being dragged through such an inter-world window would be very much like being thrown out of a jet traveling at 900 mph, and told to hang on to a trapeze bar. The first thing that would happen is that both of your shoulders would dislocate. Shoulders dislocate with 325 Newtons of force, well below the force the wind is exerting. If your shoulder is dislocated, there's a very low probability that you're going to be able to hold on to much of anything, since the major nerves traveling down the arm are being stretched a lot more than they usually are, and nerves don't like being stretched. While we're thinking about dislocations, if any part of our daredevil's legs had been twisted at all by the wind, it's very likely she could have also dislocated one or both knees or ankles.

You *could* rig yourself so that you were attached via steel cabling to something anchored on the Earth. A 3 × 3 cm cable (which is pretty industrial) will hold you fast, but you'd definitely need the assistance, because the wind, at 920 mph, is faster than the terminal velocity for a human, so it would certainly carry you along with it.

You'd also want to make sure that the Earth-side room had nothing at all which could move when subjected to a 900 mph wind. At those speeds, a 10-pound object hitting our human would inflict about 1,860 Newtons of force, which is more than enough to break a finger bone. A 5-pound object would cause a wrist fracture similar to the kind of injuries that boxers can get, regardless of how well attached you are.

There is a story about a fighter jet pilot, Captain Brian Udell, who ejected from his aircraft at 800 mph. His shoulder was dislocated, his knee was dislocated, his other ankle was broken, and all the capillaries in his face were broken from the force of the wind – his head swelled up to the size of a basketball, and his lips swelled to the point where he had a hard time moving them. He survived, but his clothes were torn to bits, and he was in the hospital and undergoing physical therapy for six months afterwards before he felt back to normal.

The wind between the Earth and the Moon is 120 mph stronger than that, so it's probably safe to assume that the capillary breakage suffered by Capt. Udell would also pose a problem for any human standing in the way of this inter-world wormhole, whether or not they were well attached to the ground. This is not a thing to try, even as a thought experiment, without really serious protective measures.

It's tempting to think that with such a wind at your back, it would be enough of a push to propel you away from the surface of the Moon, drifting forever in the space between the planets. Anything that gets blown through the window between the Earth and the Moon would have an interesting journey, but even traveling at 920 mph isn't enough to escape the gravitational pull of the Moon. Escape velocity for the Moon is 2.5 km/s and 920 mph corresponds to 0.411 km/s, six times too slow to escape. We can, however, calculate how long any object would spend flying in space before crashing down to the surface of the Moon. With our window, our hapless traveler would be flung out straight 'up' from the surface of the Moon. If we assume that any objects were sped up to wind speed, they'd have eight and a half minutes of flight (reaching a grand old height of 52 km above the surface of the Moon) before crash-landing on the Moon again at 920 miles per hour (probably destructively).*

---

* Lithobraking!

✳

The Moon is a harsh and jagged place. The scenery is rugged, and without an atmosphere, the Moon's surface is both unprotected from the hazards of the solar system and a dangerous place for humans. It is, however, our closest stepping stone in any journey outwards, and the nearest of the large objects in our solar system. In the next chapter, we'll step outwards again, and explore our Earth's cosmic siblings in the other planets, orbiting our own star.

3

THE SOLAR SYSTEM

Drifting ever outwards

Keeping planet Earth and its Moon company on their journey around the Sun are a set of planetary siblings. Orbiting the same home star, these planetary companions to our Earth travel our night skies, earning them the name 'wanderers' to the Greeks (*planetes*).

There are eight major planets: Mercury, Venus, Earth, Mars, Jupiter, Saturn, Uranus, and Neptune. Mnemonics to remember the order abound, but the International Astronomical Union's current favorite is 'My Very Educated Mother Just Served Us Nachos'★. We live on the third closest of them to the Sun (see color plate 3), and we've designated the distance between our planet and our star as a unit of distance for the entire solar system: the astronomical unit (au), which is about 93 million miles

··

★ When I was growing up, before Pluto's reclassification as a dwarf planet (see page 93), the mnemonic was 'My Very Educated Mother Just Served Us Nine Pizzas', but it has had to be reworked.

or about 150 million kilometers. So Mercury, the closest planet, orbits the Sun at 0.4 au, four-tenths the distance of the Earth from the Sun. Venus orbits around 0.7 au, and Mars at 1.5 times the distance of the Earth. The outer solar system, beyond Mars, orbits much further from our Sun, ranging from 5.2 au for Jupiter, to a little under 10 au for Saturn, and just over 19 au for Uranus. Rounding off the major planets is Neptune at a distance of 30.1 au from the Sun – 30 times more distant from the Sun than our own Earth.

In between and beyond our eight major planets there are other objects: comets, asteroids, and minor planets. The minor planet class contains things in our asteroid belt, between Mars and Jupiter, like Ceres, but also the rather beloved Kuiper Belt object Pluto. Let's explore our Earth's planetary siblings before we go any further up our cosmic family tree.

Our planetary siblings

Our solar system, which is made up of the Sun and all the planets, along with some other, smaller objects, formed roughly 4.5 billion years ago, out of a cloud of slightly rotating gas and dust. This cloud was just dense enough to start to collapse in on itself, just heavy enough to feel its own weight. The story of the origin of our planets is closely tied to the origin of the Sun itself, which sits at the very center of the solar system – before the formation of the Sun, the planets were nothing more substantial than dust and gas. The vast majority of the cloud which produced the Sun sank into the rapidly collapsing center, forming the nucleus of our solar system. Around 100 million years later, the planets began to form around the young Sun. This collapse wasn't perfect, or we'd have wound up with solitary stars and nothing left over to form planets, instead of a rich family of planetary companions for our Earth.

To get from a diffuse cloud of gas to the planets, a few things have to take place. To get the planets as we know them, we first have to make a star, and to make a star, we have to get gravity to pull a gigantic cloud of gas down on itself. This might begin by having the gas cloud disturbed, jostled into a denser state than it had been. Another possibility is that the cloud could simply have had enough time to cool, and with cooler temperatures come higher densities. This is the same principle behind hot air rising over cold air, or cold air sinking below hot air – the cold air is denser, and sinks below the less dense hot air.

Once the cloud begins to collapse downwards, gravity joins in, and the cloud will also begin to collapse inwards because of its own weight. As it does so, the cloud will also start to rotate a little. The more the cloud collapses, the more it rotates, until instead of a blobby cloud of gas, you have a relatively thin, rotating disk of gas. You can do this experiment at home if you have a chair that spins. If you start yourself turning a little bit with your arms and legs extended, and then pull your legs and arms in quickly, you'll find yourself spinning much more rapidly than you were when you started out. Pulling your arms in plays the same role that gravity played in the collapse of the early solar system.

Once we've gotten this far, our cloud of gas has converted itself into a spinning disk of gases but most of the gas has gone straight down into the middle. If there's enough there, that central core will be able to form into a star. But even with the star absorbing the vast majority of the material which collapsed, there's still a reasonable amount of gas that didn't make it in that far and is hanging around in the disk (see color plate 4). This material didn't go into the star because it was too far away to be easily pulled inwards, or was spinning too fast. This leftover gas and dust out further from the star is still relatively cold, and cold gas tends to collapse. (This is how we got the disk in the first place!) There's usually not enough mass in the rest of the disk to form another *star*, but the gas and the dust can begin to condense down and stick together, gradually forming small grainy bits of stuff – little chunks of solid material, in among the swirling gas around the star.

These new chunks of stuff orbiting around the star will crash into each other – they can either knock each other apart (meaning this growing process has to start over), bounce off of each other (meaning that the encounter did neither of them any good), or they'll glue themselves together and become one larger lump. This method of growth works until the proto-planets gain enough mass to start attracting objects through gravitational forces, which can help them gain mass even more quickly. Once a particular lump reaches a certain mass, its own gravitational weight will start making it rounder and rounder. For some objects, this process will continue until they form a fully-fledged planet.

This ideal case of catching smaller lumps of material is rarely a smooth process. In the earliest stages of building up a planet, the speed with which it can gain mass is largely dependent on how much nearby matter there is to be wafted in the right direction to clump on to our object. Ultimately, this means that some infant planets will grow more rapidly than others, if they find themselves in a region that has a lot of material with which to grow. In turn, this means that the early solar system was a very hectic place, with swarms of little 'planetesimals' of various sizes. Some of these planetesimals will unavoidably have been absorbed into one of their larger neighbors, building up the largest planets to the sizes we see today.

The gas disk of the early solar system also determined the direction the planets were going to orbit around the central star. None of them were going to suddenly come to a stop and reverse course without something very violently turning the whole thing around, which would be hard to arrange. The planets in a typical solar system should also be pretty close to orbiting in a single plane around their star – that plane effectively shows us where the gas disk once was. Our solar system is no exception to these expectations. Every planet goes around the Sun in the same direction, and the main eight planets are all aligned in a very thin plane. We can see this ourselves in the night sky: all the planets always appear along a single arc in the sky, which, coincidentally, the Moon also follows, named the ecliptic.

Planetary spin

The same kind of spin-up that happened when the cloud of gas collapsed downwards into a disk continues when you go from a disk of material to a set of planets. Planets are more dense again than the disk of the early solar system, so we expect most planets to continue the pattern of spinning. And indeed, in our solar system, which is our easiest to observe set of planets, all the major planets are spinning around their own internal axis.

We're well acquainted with the rotation of the Earth, of course, because that's what creates our days. We all know it takes 24 hours for the Earth to rotate, but 24 hours isn't the rule within our solar system. In fact, of all the other planets, only Mars rotates at a similar speed. Mars completes one rotation in 24 hours and 40 minutes; nearly identical to our home planet. Venus, meanwhile, rotates much slower than the Earth – one rotation takes 243 days and 26 minutes; this makes it the slowest rotator in our entire solar system. Mercury comes in second slowest, rotating once in 58 days, 15 hours, and 30 minutes. The gas giants, Jupiter, Saturn, Uranus, and Neptune, all seem to have similar rotation speeds, all faster than any of the inner, rocky worlds. Jupiter rotates once every 9 hours and 55 minutes, and Saturn is close behind, at 10 hours and 36 minutes. Uranus rotates once every 17 hours and 14 minutes, and Neptune once every 16 hours and 6 minutes.

With two exceptions, every major planet in our solar system rotates in the same direction – the same direction as the Earth. If we had a bird's-eye view of our solar system, where we'd flown into space 'up' via the North Pole and looked back down, most of the planets would be rotating counterclockwise – or from the west towards the east. You can usually remember which way the Earth is rotating by thinking about the time zones: the further east you go, the later it is – they're pushed towards daylight sooner than the west. This consistency in spin direction is expected, again because all the planets are shrinking down out of a disk of gas that was itself spinning. It would be a violent, difficult task to reverse that spin.

But it's not impossible, and our solar system has a few examples that show it can happen. Venus and Uranus both spin in an unusual direction. Uranus rotates 90 degrees off from everything else. If you consider the plane of all the planets' orbits around the Sun as a flat surface, most planets spin as though they were a coin spun on its edge, flicked counterclockwise. Uranus, on the other hand, spins like a bead rolled along the ground, instead of spinning vertically. Venus is even odder – it spins clockwise. As far as we can tell, this means Venus is upside down.

Both planets should have *formed* with a rotation aligned with every other planet in the solar system, so to get to their current upside-down and sideways state, something must have changed them. In fact, that's probably what happened. The early solar system didn't just form eight planets, it formed many more, and collisions between those planets under construction was a relatively common occurrence. A particularly bad collision or series of collisions could have caused such an energetic punch to the growing Venus or Uranus that the planet's entire orientation was shifted, leaving the planet out of sync with the rest of the solar system.

A stable siblingdom

Fortunately for life on planet Earth, the kinds of collisions which might have tilted Uranus and Venus are a thing of the distant past. As material was collected out of the disk of gas and dust, and into a set of small objects that might eventually become planets, the early solar system was also cooling down. Unlike a gas cloud, which can retain heat a little better, individual balls of rock are not so good at keeping the entire solar system warm.

At some point, the solar system had cooled to such a point that it became difficult for the newly formed rocks to stay hot. That heat had allowed these rocks to stick together more easily, instead of bouncing or shattering, as the rocky material was pliable, like molten glass. Without this stickiness, the growth of most objects stopped. To keep growing from this point, the larger objects had to collide with others – but many of these smaller pieces have remained just as they were at the end of this phase. Asteroids and comets are among these remnants of the formation of the solar system, and date back some 4.5 billion years.

Because the solar system has had billions of years since those earliest days, most of the large objects have either already had the collisions they're going to have, destroying themselves in the process, or they've survived such encounters and have found a stable place to orbit the Sun, free of disturbance. This disturbance-free orbiting is such a feature of the major planets in our solar system that it has now been folded into the definition of a major planet – no other similarly sized objects should cross the orbit of a major planet.

There are a few objects that seem to have done okay in building up mass to the point of making themselves round, and also managed to avoid being destroyed by colliding with a larger planet, but then failed to continue gaining mass to the point where there weren't any other objects in their orbit. These are what we now call dwarf planets, or minor planets. They made it partway to being a planet, but didn't quite get all the way there. (See the section on Pluto, page 93.)

The pieces of our earliest solar system which never quite made it into a full-sized planet, the asteroids, can be scientifically fascinating. These objects serve as time capsules to the formation of the early solar system, telling us what it was made of and how fast it was cooling down, among other things.

In order to get a smaller object to stick to a young, growing planet and gain its mass, the force of the collision is the deciding factor. This is controlled by two things – the speed at which the two objects collide, and the mass of the two objects. If the two objects are the same mass, but collide at very high speed, they both may shatter. This kind of head-on collision is what we think happened to create an X-shaped asteroid observed by Hubble in 2010 (see color plate 5). You can just as easily shatter a smaller object by running it into a much larger object. This is the same as breaking a rock by dropping it on the ground from a height.

The shards of shattered objects from the early solar system can end up wandering the solar system, and many of them are found in the main asteroid belt between Mars and Jupiter. There's another reservoir of asteroids and comets which sits out beyond the planet Neptune. We encounter the smallest pieces of the early solar system with every meteor shower, and very occasionally, one larger piece will survive its passage down through the atmosphere and land somewhere on Earth. From the analysis of these pieces, we've learned quite a lot about the early solar system, and what metals were present there.

Learning from meteorites

The meteorites found on Earth are usually made of some combination of rock and/or iron. With the infinite inventiveness of astronomers, we duly called them stony, iron, or stony-iron, if they're a relatively even blend of the two. Stony meteorites can (and often do) contain some small fraction of iron, but only those meteorites which are almost entirely metal are classified as iron meteorites. The vast majority – over 90% – of meteorites that fall to Earth are stony.

The problem with spotting stony meteorites is that they look like rocks, and there are a lot of rocks on our parent planet. The only place where it's easier to spot this type of meteorite is where you wouldn't expect to find many rocks naturally. The best place to look for stony meteorites is therefore the glaciers of Antarctica. Glaciers do not normally have rocks on top of them, and the ice there doesn't move very rapidly, so any meteorites that fall there will both stand out against a bright white background and stay put for quite a long time. These meteorites are some of the oldest untouched rocks in the solar system, and so are profoundly interesting to scientists as conveniently delivered time capsules from the very beginnings of that system.

There's one other interesting set of stony meteorites – we have assembled a collection of about 30 meteorites which have come to Earth from Mars. These are chunks of the surface of Mars that were flung away after a larger object hit the red planet. These pieces then wandered the solar system until they encountered the Earth. These Martian meteorites tend to be much younger than the rest, and are composed of a different set of elements and minerals, so we can tell them apart from the older stony meteorites without too much trouble.

Iron meteorites are much easier to spot than the stony ones. This is partly because it's relatively unusual to find weathered chunks of iron sitting on the ground, and so it's easier to recognize that they are out of place. As with the stony meteorites, they are most easily found in deserts and in the glaciers of Antarctica, where they are likely to stand out more from their background, but they can also be found in other places. These metal meteorites are also less affected by erosion, so they can be identified for a much longer period of time after they fall.

The metal in these meteorites is not entirely iron, though that's a large fraction of it – the remainder is nickel. This blend of metals is part of what makes them unique. Another part is that they tend to have long crystals of metal within them which can be sliced and polished to show the crystals off. (Without this extra treatment, they look much less dramatic.)

Opportunity, one of our Mars rovers, discovered a basketball-sized iron meteorite on Mars in 2005, which was particularly exciting, because it was the first time we'd found a meteorite on a planet other than our own. Its irregular, pitted, and lumpy look is much more typical of the appearance of a nickel–iron meteorite.

Iron meteorites and stony meteorites also behave a little differently as they come through the atmosphere. The stony ones tend to be more fragile against the force of the atmosphere, and as a result they can shatter more easily before they hit the ground. Iron meteorites are more resilient. The heat of re-entry doesn't usually stress the metal to its fracture point, but it can certainly reach that point for stone.

The Chelyabinsk meteor that exploded over Russia in 2013 also left behind some pieces for us to examine; the largest of these were extracted out of a lake. Because it exploded so dramatically in mid-air, it had been suspected to be a stony type object, and the pieces confirmed that hunch.

Both iron and stony meteorites give us a look at the early solar system, and help us to understand how the planets formed, when, and out of what materials. As we grow our meteorite collections, scientists are able to understand what kinds of variation might have been present in that earliest disk of infant planets.

Heavy metals in the solar system

Asteroids and planets, unlike the stars, have no fusion occurring at their cores and thus cannot make any new elements (see page 125). This means that all the metals present on the Earth and in all the asteroids were already there, in the gas cloud that our star formed out of, before the planets began to form.

This is interesting, because by all current models our universe began with a pretty limited store of elements. Hydrogen, a little helium, and a tiny bit of lithium were the only elements to be had. They're the three lightest elements in the periodic table, and so every element heavier than hydrogen and helium has been added later into the universal mixture.

These elements have been added by stars. We'll go into more detail on how the stars can build up so many more elements in the next chapter, but the key point for now is that without the presence of stars in the area before our solar system had formed, all the carbon, silver, and gold we find on our planet would not exist.

The iron and nickel that make up both the core of planet Earth and the metallic meteorites are both formed in the very centers of large stars, at the very end of their lifetimes. All of the stars which are capable of creating nickel and iron go through a supernova explosion at the end of their lives, triggered when nothing remains in the core of the star that can be burned. (More on this in Chapter 4.) The burn that creates nickel is the very last one possible, so a supernova is imminent once this process starts.

Supernovae are pretty sizable explosions and spread some of the elements they've created into the space that surrounds them. Over many generations of stars, the small amounts of more complex elements that have been created and dispersed build up, to the point that any given gas cloud has probably been filtered through a few stars at some point in the past. So it was with the cloud of gas which became our solar system – the gas had already been recycled through a few rounds of stars, which had filled the gas with a good amount of nickel, iron, and all the other elements we now spot on our planet.

THOUGHT EXPERIMENT:
A hypothetical interplanetary delivery system

Since we know that pieces of the early solar system are regularly smashing into planet Earth, and we know that they've also smashed into Earth's sibling planets, questions have come up as to whether the meteorites which make it to the surface could carry either life, or the materials for life, buried deep within them. Another

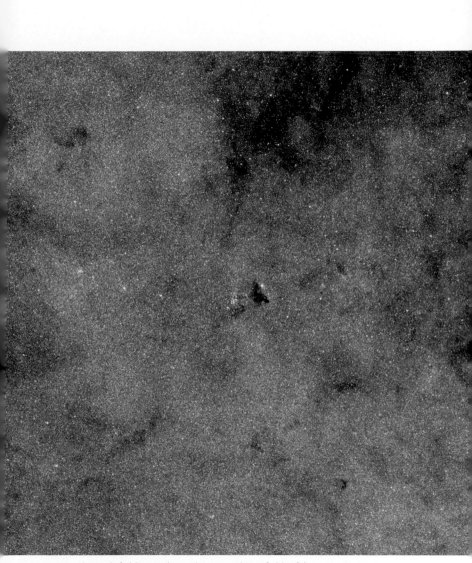

1. This wide-field view shows the very rich star fields of the Large Sagittarius Star Cloud and the cluster NGC 6520 and the neighboring dark cloud Barnard 86. It was created from images from the Digitized Sky Survey 2. (ESO/Digitized Sky Survey 2; Davide De Martin, CC BY 4.0)

2. Composed of gas and dust, the pictured pillar resides in a stellar nursery called the Carina Nebula, located 7,500 light years away in the southern constellation of Carina. Taken in visible light, the composite image was made from filters that isolate emission from iron, magnesium, oxygen, hydrogen and sulfur.
(NASA, ESA, Hubble SM4 ERO Team)

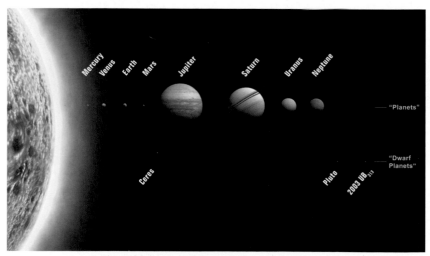

3. The eight planets and the new solar system designations.
(International Astronomical Union)

4. The protoplanetary disk surrounding the young star HL Tauri. This is the sharpest image ever taken by ALMA (the Atacama Large Millimeter/submillimeter Array telescope in the Chilean Andes) – sharper than is routinely achieved in visible light with the NASA/ESA Hubble Space Telescope. These new ALMA observations reveal substructures within the disk that have never been seen before and even show the possible positions of planets forming in the dark patches within the system.
(ALMA (ESO/NAOJ/NRAO))

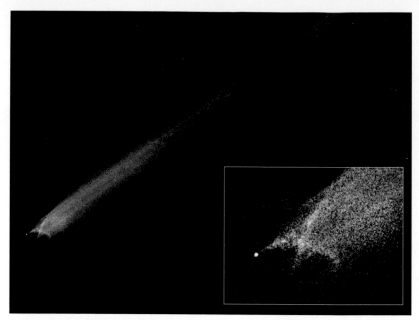

5. This is a NASA/ESA Hubble Space Telescope picture of a comet-like object called P/2010 A2, which was first discovered by the LINEAR (Lincoln Near-Earth Asteroid Research program) sky survey on January 6, 2010. It shows a bizarre X-pattern of filamentary structures near the point-like nucleus of the object and trailing streamers of dust. The inset picture shows a complex structure that suggests the object is not a comet but instead the product of a head-on collision between two asteroids traveling five times faster than a rifle bullet.
(NASA, ESA and D. Jewitt (UCLA))

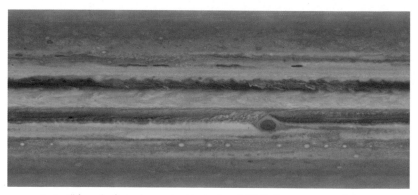

6. A still from the first color movie of Jupiter from NASA's Cassini spacecraft shows what it would look like to peel the entire globe of Jupiter, stretch it out on a wall into the form of a rectangular map, and watch its atmosphere. The smallest visible features at the equator are about 370 miles (600 kilometers) across. In a map of this nature, the most extreme northern and southern latitudes are unnaturally stretched out.
(NASA/JPL/University of Arizona)

7. The largest canyon in the solar system cuts a wide swath across the face of Mars. Named Valles Marineris, the grand valley extends over 2,400 miles long, 125 miles across, and 4 miles deep. By comparison, the Grand Canyon in Arizona, USA is 275 miles long, 18 miles wide, and 1.1 miles deep. The above mosaic was created from over 100 images of Mars taken by Viking Orbiters in the 1970s.
(NASA)

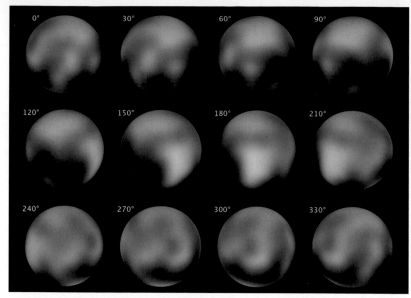

8. This was the most detailed view to date of the entire surface of the dwarf planet Pluto, as constructed from multiple NASA Hubble Space Telescope photographs taken from 2002 to 2003. Pluto is so small and distant that the task of resolving the surface is as challenging as trying to see the markings on a soccer ball 40 miles away.
(NASA, ESA, and M. Buie (Southwest Research Institute); photo no. STScI-PR10-06a)

9. Pluto nearly fills the frame in this image from the Long Range Reconnaissance Imager (LORRI) aboard NASA's New Horizons spacecraft, taken on July 13, 2015 when the spacecraft was 476,000 miles (768,000 kilometers) from the surface. This view is dominated by the large, bright feature informally named the 'heart', which measures approximately 1,000 miles (1,600 kilometers) across. Even at this resolution, much of the heart's interior appears remarkably featureless – possibly a sign of ongoing geologic processes. (NASA/APL/SwRI)

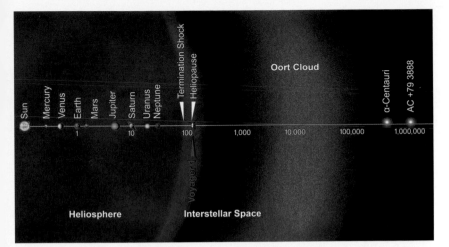

10. This artist's concept puts solar system distances in perspective. The scale bar is in astronomical units (au), with each set distance beyond 1 au representing ten times the previous distance. One au is the distance from the Sun to the Earth, which is about 93 million miles or 150 million kilometers. Neptune, the most distant planet from the Sun, is about 30 au. The approximate location of Voyager 1 is shown in red. (NASA/JPL-Caltech)

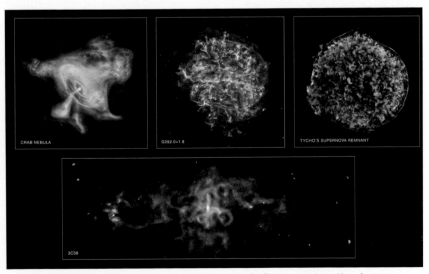

11. Four images of supernova remnants dramatically illustrate NASA's Chandra X-ray Observatory's unique ability to explore high-energy processes in the cosmos. (NASA/CXC/SAO)

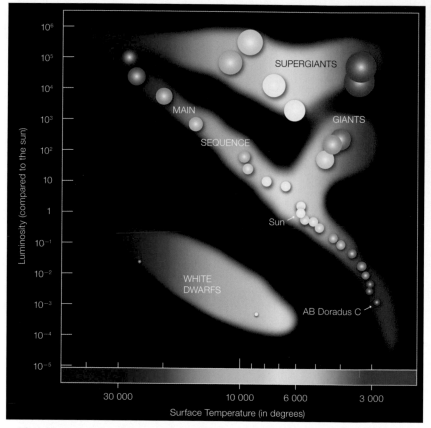

12. A Hertzsprung-Russell diagram showing many well known stars in the Milky Way galaxy. (ESO CC BY 4.0)

option for this kind of delivery system is to remember that the Earth has been on the receiving end of some Martian rocks – so perhaps once one planet develops complex organic materials, it can share these with its siblings via meteorites.

These ideas are really only suggestions – there's no proof that this is how the solar system has done it, or if such a delivery system is even necessary. But if you wanted to naturally transport life around a solar system, how could you do it?

The hypothesis that suggests that either bacterial life or the building blocks for life could be transported via comets or meteorites is called 'panspermia'. It proposes that once life arises somewhere in the universe, a large impact upon that planet could create such an explosion that pieces of rock carrying bacteria are flung into space, where they cruise about until crash-landing on another life-friendly planet.

The real unknown is whether or not bacterial life is hardy enough to survive this process, including the blasting off the original planet and the landing on the new planet. If it can, then this might be one way to get bacterial life from a planet to its siblings within a solar system, however unlikely it might be for those pieces of rock to be flung in exactly the right directions.

Scientists have been testing whether bacteria can survive being exposed to the vacuum of space at all.

A few microbes from the town of Beer* in southwest England managed to partially survive a 533-day stint on the outside of the International Space Station. This is impressive, but you should take note that it was a partial survival of one strain of bacteria, on a relatively short exposure time for a between-planets unplanned journey. There's no guarantee that a rock blasted off the surface of another planet will be covered with a particularly space-hardy breed of microbe, nor that the meteoroids will arrive at another planet in a short period of time.

It might instead be easier to transport not the live bacteria, but the organic molecules† that are required to build them. Amino acids are one such set of complex organic molecules, and are often called the building blocks of life. If a comet or meteorite can bring amino acids to a planet, it might help to jump-start the development of life on that planet by skipping the steps required to build those molecules from scratch. These molecules don't need to have formed on the surface of another life-bearing planet – surprisingly complex molecules can

---

* Why Beer? The sea cliffs near Beer were found to have a set of microbes with extra thick cell walls, so it was thought they might survive better than the average microbe. The experiment just sent up a chunk of the cliff rock and attached it to the outside of the ISS.

† A molecule is a combination of atoms, bound together, typically by sharing electrons (see page 108). The simplest molecules are two atoms tied together, but much more complex molecules make up all life.

form in the gas between stars in our galaxy. Because these simpler building blocks may be more widely available than planets with life that are also being bombarded with impacts, it may be easier to distribute these molecules than to distribute entire bacteria.

Until we have more concrete proof of either of these processes happening out there, they will remain intriguing possibilities, but we're working towards testing each of the steps individually to get a sense of how plausible all of the steps together might be.

All the planetary siblings to our Earth are fascinating worlds in their own right, but a few of them appear to have caught the public imagination more than others. To that end, let's take a closer look at a couple of our neighbors, beginning with Mars.

The chasms and volcanoes of Mars

We humans have sent a lot of robotic explorers to Mars, and with a number of rovers on its surface, we have an incredibly detailed view of what the red planet looks like up close. The rovers have gathered a great body of evidence to tell us that there was very likely a lot of warm, life-friendly water on the surface of Mars for quite some time. The Opportunity rover in 2011 found a huge streak of gypsum sticking out of the surface, along with hematite 'blueberries' – both signs of a wetter past on Mars. There's nothing as far as we can tell that might have entirely prevented life from arising on Mars a billion years ago, but equally we haven't found any definitive evidence that there was life, either.

However, the current environment on Mars is very different from its watery past – it's now a cold and cancer-inducing place for humans. There's still water there, but it is usually in the form of ice. The polar ice caps on Mars are both substantial in size and mostly water ice. These grow and shrink with the seasons, just like our own ice caps. Mars has such dramatic dust storms that the entire planet can be engulfed in them, hiding the features of the surface from our view.

And as wonderful as the images being returned from the rovers exploring Mars are, they present a very biased view of what the surface looks like. The rovers have been sent to rather flat ground, which gives relatively familiar-looking horizons, instead of something jagged and foreign. There's an important reason for this: we can't land rovers on very rocky ground or at high elevation. We need landing sites to be at low elevation so that there's a lot of atmosphere to help slow the rover down, and we want the ground to be flat so we don't accidentally drop a very expensive machine on a boulder. These restrictions on our landers mean that we miss out on the most dramatic vistas Mars has to offer.

Valles Marineris is one of these places, and it makes the Earth's Grand Canyon look unimpressive. At more than four times as deep, and five times as long, Valles Marineris would be challenging territory even for humans. Trying to land something the size of a large car on the jagged ground there is impossible for now. Water very likely flowed through it, but it's unclear whether it was formed by the water itself, or whether it is a more fundamental crack in the planet's crust. In either case, it is the solar system's largest canyon (see color plate 7).

On top of having the largest canyon, Mars also has the solar system's biggest volcano: Olympus Mons, one of a series of extinct Martian volcanoes. It's 14 miles high. That makes it three times higher than Mount Everest, and twice as big as Mauna Loa in Hawai'i, if you start counting from the ocean floor (see image below). Olympus Mons has cliffs several miles high right at the base, which will make for an incredible climb whenever we get people over to Mars, but is a rather severe impediment to getting rovers onto the volcano. (This particular problem seems to be unique to Olympus Mons – the smaller volcanoes don't have the same kind of cliffs.)

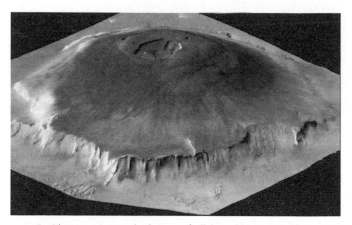

5. Olympus Mons is the largest of all the volcanoes on Mars, as well as the largest known volcano in the solar system. It is a shield volcano about 370 miles (600 km) from edge to edge, and would fill the entire state of Arizona. (NASA/JPL)

Hawai'i is actually a pretty good comparison – the two are the same style of volcano, but where Hawai'i has moved in relation to the hot spot beneath it, which keeps the volcanoes from getting too huge, Olympus Mons stayed put. Mars has never had any plate tectonics,* so there was

* Segments of the Earth's crust drift relative to each other, colliding and spreading apart at the various boundaries between segments. This is known as plate tectonics. Where individual segments (plates) collide, mountains and often volcanoes are built. Where they separate, deep rifts can arise. Planets without plate tectonics must build mountains, valleys, and volcanoes through other means.

nothing to move the volcano, and it could grow to a tremendous size.

The presence of volcanoes on the surface of Mars brings up another intriguing possibility. On Earth, volcanoes serve to dredge up some of the heavier metals which would have sunk to the center of the planet while everything was still molten, and before the crust had solidified. On Earth, these upwellings are responsible for lifting precious metals such as gold back up to the surface, as it is dissolved in with the rock and volcanic lava. We can also get some additional gold at the surface from impacts from asteroids, but at least on Earth, it seems that volcanic regions are better places to find gold.

Mars being a sibling to our Earth, and forming out of the same mixture of materials, it should also have roughly the same fraction of gold as we do, and it has the volcanoes which could bring this metal back up to the surface. While the Earth's volcanoes are active *now*, Mars' were active a billion years ago, but otherwise it seems reasonable to think that this redepositing of gold on or near the surface may well have also happened on Mars.

It's not just gold that volcanoes bring to our Earth – they also produce geodes. Geodes are hollowed-out bits of rock, with crystal formations growing on the inside. The type of lava which flows smoothly (instead of exploding) frequently forms small bubbles. Sometimes these bubbles pop, but sometimes they harden in place. If the bubble survives, over time, water may be able to seep inside, depositing minerals, which can form crystals. We know now that Mars had plenty of water at one point, so this process may also have been possible on Mars.

Agate also has a volcanic origin. Agate forms when material full of silica (the main ingredient in white sand and glass) slowly fills or partially fills an irregular hole, perhaps formed by a gas bubble in lava. As the material fills in the hole, it creates the bands you see in some agate, like a tree produces bands as it grows.

The major stumbling block in suggesting that there *might* be geodes, agate and gold on the surface of Mars is simply that we'd need to go near the volcanoes to find them, and as we've said, that's a particularly dangerous place to try and put a rover. It has proven difficult enough to get any kind of machine to the surface of Mars, so making a landing extra precarious has not been at the top of anyone's priority list. Until we can figure out a safe way to explore the volcanic regions of Mars, either with rovers or a human crew, the presence of gold, agates and geodes will remain simply plausible – and the prospect of Martian gold jewelry a far-off possibility.

Let's not contaminate Mars

With all the hints of a wetter, possibly habitable past on Mars, as we go to explore the red planet, it makes very good sense to try our very best to avoid contaminating the surface of the planet with our rovers. And indeed, contamination of other worlds is a major source of concern, particularly when we're going someplace that might be habitable for some form of life. All spacecraft (and this includes satellites, telescopes, and rovers) have all of their pieces go through some pretty serious decontamination before launch, and most are assembled in a very high-test clean room. (The last time I was near a clean room of this caliber, we weren't even allowed in the corridor that eventually led to the clean room, to try to keep airborne particles to a minimum.)

Part of this clean assembly is for the good of the craft itself. Discovering that, somehow, dust had settled all over the mirror of your brand-new telescope would be a nightmare scenario, meaning that it's seriously underperforming, to the detriment of all the science that it might have been able to do. And you certainly don't want grit making its way into the moving parts, which could jam their operation. If the part that needs to move is critical to the success of the mission (say, for instance, it's your communications antenna), a stuck part could derail the entire project.

6. This self-portrait of NASA's Mars rover Curiosity combines dozens of exposures taken by the rover's Mars Hand Lens Imager (MAHLI) in 2013. (NASA/JPL-Caltech/MSSS)

If your spacecraft has a gas chromatograph or a mass spectrometer on it, both of which are instruments designed to sample the molecular properties of a gas, then you have to be extra careful. If you have such a sensitive metric of the chemical composition of air or vaporized soil, any kind of contamination left from Earth on the tray you're using to hold the gas or dust will contaminate your measurement, and you'll wind up measuring the contaminant instead of what you really want it to be sampling (i.e., Mars). This has actually been an issue in the past: the Viking landers on Mars in 1976 made this kind of measurement of the soil and found some unusual chemicals in it. Unfortunately for the soil measurement, it was found that you could get this reading either by having Mars be doing some interesting chemistry, or if there was still some residual cleanser on the soil tray.

If we add in some later measurements by the Phoenix lander in 2008, it's possible that the 'interesting chemistry' interpretation is actually the correct one, but until some more measurements are made on the exact isotopes* of the

....

* Not all copies of an element have the same number of neutrons (see page 128) in their cores. These differently constructed atoms are called isotopes of each other. They remain the same element, because they contain the same number of protons (see page 127) in the core of the atom, but the mass of the atom changes.

7. In NASA's Viking project, two identical spacecraft, each consisting of a lander and an orbiter, flew together and entered Mars orbit; the landers then separated and descended to the planet's surface. This photo shows a test version of the landers in the original 'Mars Yard' built at NASA's Jet Propulsion Laboratory in 1975. (NASA/JPL-Caltech/University of Arizona)

atoms involved in the reading, it'll be hard to rule out the cleanser contamination explanation. The Curiosity rover, on the other hand, seems to have found genuine complex molecules from Mars air and vaporized rock, but the scientists working on the team could only say this after determining that the results of the experiment on different rocks were sufficiently dissimilar to rule out contamination. If you thought the signal you were getting was from a single contaminant on your tray, it would produce a similar signal no matter which rocks you were inspecting, so finding different signals from different rocks means that it's probably something attached to the rocks, and not to your measuring device.

Landers so far have been placed on parts of Mars and in other places in the solar system that are pretty dry and reasonably flat. The dry tends to go along with the flat; by and large our best observations of liquid material flowing on Mars have been along steep slopes, where it is too dangerous to land. Avoiding those places for the purposes of avoiding contaminating any water reservoirs is done out of an abundance of caution. It's hard to *guarantee* that we don't have any cleanser products or other contaminants left on our rovers, in spite of our best efforts to send as clean a craft as possible. And if there is some form of extreme life out there on Mars, we'd like to not kill it accidentally.

Jupiter's stripes

From Mars let's move out one planet, to Jupiter. Jupiter is the largest planet in our solar system – a tremendous assembly of gas, likely wrapped around a relatively small, rocky core (though recent data from the Juno spacecraft has indicated that the center of Jupiter is fuzzier than expected).

Jupiter is classed as a gas giant, as are the other outer planets, and is more massive than all the rest of the planets in the solar system combined. It's also host to an enormous, hundreds-of-years-long storm, which we know as the Great Red Spot. And crossing the planet, from top to bottom, are cloud bands, alternating in color (see color plate 6).

Jupiter's atmosphere is doing some pretty weird things, and we don't fully understand all of what's going on. We know that the light stripes and dark stripes are made of slightly different gases. At the boundaries of these stripes are narrow jets of high wind, which push the nearby atmosphere around with it; the bands themselves are relatively stable.

The light stripes seem to be made of cold gas which is coming up towards the surface of the atmosphere, and the dark stripes are warmer gas, sinking down towards the center of the planet. The light ones are light because there's a lot of ammonia in the upper atmosphere of Jupiter, and as it cools, it forms pale clouds, like the clouds in our own sky. If the gas warms up, the clouds will disappear, and what we're seeing as dark bands are actually a deeper, darker layer of clouds. Jupiter's still doing something unusual with those bands, because the Sun also has gas rising and falling with different temperatures, and the Sun doesn't have any such stripes.

At a very basic level, Jupiter's stripes are present because there are jets of wind running around the planet. The jets form a boundary for the gas, and the gas is easily redirected along the path of the wind. The jets alternate in direction as you go from the equator to the poles of Jupiter, which means the atmosphere is being pushed in different directions, depending on where you are on the planet. At the edges, eddies form, the way swirls form if you push your hand backwards against the flow of water. The wind jets mark the edges of the bands, and each stripe moves in a different direction. The part that we don't yet understand is *why* those jets exist in the first place.

Broadly, there are two main ideas. One is that this is turbulence at the surface level, like clouds in the upper atmosphere of the Earth. Perhaps there was some turbulence – a little bit of bumpy air – and it ran into another patch. As patches of turbulence catch up to each other, they can combine in what's called a cascade. If there's a constant source of the little eddies of turbulence, then you can maintain bigger turbulence (like the wind jets) just by tossing the little ones together. But while this method can create the jets, it's not great at keeping them stable.

Since Jupiter is rotating relatively quickly (a Jovian day is just under ten hours long), and seems to be all gas, the gas could alternatively form cylindrical shells of material that rotate in different directions as you go out away from the center of the rotation. Planets are not cylinders, but if you start with a cylinder, and carve into it at top and bottom, you can create a sphere. So these rotating bands might appear as the surface of Jupiter cuts into different layers, as you get closer to the poles. This is similar to the way that a sharpened wood pencil has a stripe of pencil lead, a stripe of the wood that's been cut into, and the outer layer of the pencil that wasn't touched, but where each layer of the pencil is rotating. This is also not a perfect solution, because generally it doesn't form enough bands to match the Jupiter we see.

Complicating the whole mess is the Great Red Spot, nestled in between these bands of high winds. The Great Red Spot is more like a hurricane than anything else, but with reliable observations of this storm spanning hundreds of years (since the 1800s), and at several times the size of our entire planet, it's both the longest-lived and the most physically expansive storm we know of. It's also a very perplexing storm, and explanations of its longevity and its size are hampered by how different Jupiter is from our Earth. A planet with no solid surfaces under its clouds may well be able to maintain hurricane-like storms for so long, but even the color of the Great Red Spot is not very well understood. *Something* must be happening with the gases which make up the storm to produce such a vivid color, but exactly what that is remains a bit of a mystery.

Since neither of the explanations for the stripes really sets out a comprehensive solution to the whys of Jupiter's atmosphere, for the moment we have to stick with observing the banded clouds and the jets that drive them, in hopes that understanding the details of their behavior will hint towards one answer over another. The Juno probe, which arrived in orbit in July 2016, is helping to answer these questions as it orbits Jupiter.

THOUGHT EXPERIMENT:
Jovian night skies

Jupiter's thick cloud layers may be puzzling, but because the planet is mostly gas, the Earth is actually considerably more dense on average than Jupiter is. Jupiter's gassy atmosphere means that on average it's only a quarter as dense as the Earth, even though it is made of more than 300 times the Earth's mass. The combination of having way more material than the Earth, and being less densely packed, means Jupiter takes up a lot more physical space than the Earth does. In terms of volume, Jupiter is more than 1,300 times larger.

With a planet that large, it may come as a surprise to learn that the number of stars in the night sky visible from the top of Jupiter's atmosphere should be the same as the number you could observe in the dark skies of our parent planet.

On Earth, we see a fixed number of stars at night because there is, at best, only 180 degrees of sky to look at. Of a full circle, half falls below our feet, and half is sky. The ground of our planet appears flat to us because the curve of the Earth's sphere is so gentle. Our half-circle of sky is what falls above our local, mostly flat, surface. As the Earth turns, our personal patch of surface points towards different stars in the sky, which causes some stars to rise and others to set in our visible sky. As you move north or south on the planet, the position of the stars shifts in the sky, again because you're pointed at a different patch of stars. Most notably (in the northern hemisphere), the North Star will appear higher or lower in the sky as you move north or south respectively. If you moved from the North Pole to the South Pole, the stars you would see would be *entirely* different, as you'd have no overlap between the 180 degrees of sky you're pointed towards at the North Pole, and the 180 degrees of sky at the South Pole.

Going from Earth to the much larger Jupiter, the only change is that the curve of Jupiter's sphere is even gentler than the Earth's. The atmosphere below you would still locally look flat, meaning that you'd still have 180 degrees of sky and 180 degrees of planet under your feet. However, because the sphere of the planet is so much larger, it takes a long distance to change your angle on the stars, so you'd have to travel much further to notice that the North Star had moved.

The only way you'd be able to see *more* stars at night, all at once, is if you were on such a small planet or moon that you could see the curve of the planet dropping away from you. Standing on these objects would feel like being on top of a planet-sized mountain. Here you would have more than 180 degrees of sky, because your horizon would always be a circle below you. You would be able to see stars up, horizontally, and at an angle down beyond your feet. You'd also be able to change your night sky very quickly, just by taking a few steps sideways, and you'd always be able to see more stars at one time, without the motion. Deimos, one of Mars' small moons, which is only about 8 miles (or 12 kilometers) across, might work.

Pluto the misfit

In 2006, the International Astronomical Union (IAU) made a change to the set of criteria an astronomical object would have to clear in order to be considered a major planet, and Pluto was reclassified into the dwarf planet category. Pluto is now part of an ever-growing group of dwarf planets, but the decision to reclassify Pluto ruffled a lot more feathers than one might have expected, for a small sibling to our Earth that, at the time, we knew precious little about. Now, after New Horizons' flyby in 2015, we know more about Pluto than ever before, and it is undeniably a fascinating little world.

But the truth is, Pluto has never fit in very well with the other eight planets. The orbits of all the major planets around the Sun lie very flat, compressed into synchronicity. The orbit of Pluto in comparison is wildly askew. It's tilted by 17 degrees relative to the rest of them. Also unlike the other planets, which travel around the Sun in almost perfect circles, Pluto's orbit is extremely elongated. Pluto is also quite small. It's big enough at 1,475 miles

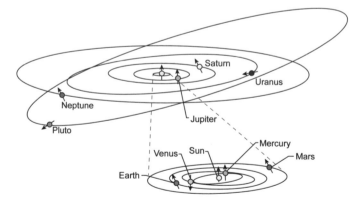

8. A drawing of the solar system showing Pluto's tilted orbit. Pluto's orbital path angles 17 degrees above the line, or plane, where the eight planets orbit. (NASA)

(2,374 kilometers) across to have compressed itself into being a sphere (compared to the lighter and more irregularly-shaped asteroids), but it has a moon in Charon that is almost as big as itself. None of the inner eight planets behave this way.

While we had noted these irregularities with Pluto's behavior, for a while it seemed to be a unique little world, surrounded by much smaller objects than itself. It seemed sensible to leave Pluto as a planet as long as it merely stayed unusual for an object so far out from the Sun. This is where Pluto started running into problems – we began to realize that it wasn't unique. Our ability to detect Pluto-sized objects got a lot better, and all of a sudden we had found a handful of other worlds. Most notably, there was Eris, which was even further out from the Sun than Pluto, and also bigger.

Now a decision had to be made with these new discoveries. Either we allowed Eris in to the solar system as the tenth planet (if Pluto was a planet, Eris surely was), or we would have to come up with a clearer definition of what a planet was. The problem with adding Eris and keeping Pluto as planets was primarily that Eris' discovery was proof that Pluto was not unique, it was merely the first in a class of objects we'd been unable to detect so far. That meant that we'd be constantly adding new planets to the list of siblings that make up our planetary family as more and more of them were discovered.

So the IAU decided to impose a set of three criteria that any object in the solar system must pass in order to qualify as a planet. They were the following:

- A planet must orbit the Sun, and not another, smaller object. (It can't be a moon.)
- It must also be sufficiently massive to have compressed itself into a sphere, which excludes all the asteroids and funny-shaped comets.
- Finally, it should have cleared the area around its orbit of other objects. This means that there shouldn't be any other worlds of a similar size anywhere along the orbit of your planet.

Earth doesn't have any other Earth-sized things sharing our orbit, but Pluto failed that last criterion – there *are* other Pluto-sized objects along its orbit. This meant that Pluto and Eris were redesignated as dwarf planets. This classification distinguished them from the irregularly shaped objects that orbit the Sun (like the asteroids), but swept them out of the main planet category.

The decision to reclassify Pluto was a conscious choice to make our definitions more consistent. Our understanding of the solar system will gradually become more and more complex as time goes on, so it makes sense to let our classifications reflect the complexity of our planetary family.

Pluto has mountains

While Pluto was being reassigned to the dwarf planet side of the family, very little was known about what it actually looked like. This all changed as soon as the New Horizons spacecraft made it to Pluto, after traveling for a decade to get there.

Before New Horizons, we could tell you that Pluto seemed to be made of a roughly even mixture of rock and ice, and that it had an abundance of moons. In addition to its large companion in Charon, which is 10% of Pluto's mass (this is huge for a moon), Pluto has four other moons, dubbed Nix, Hydra, Kerberos, and Styx.

The only real images of Pluto to be had were from a combination of Earth-based observatories and some space-based, like Hubble (see color plate 8). Pluto was still too far away and too small to do much with – the highest resolution images that Hubble could return were still very fuzzy. The resolution was so poor that we could really only tell that there were variations in brightness and color, instead of it being more consistent all over, like many of the icy moons of Jupiter or Saturn.

Sending New Horizons was a long-term project, and a risky one. On top of the ten-year journey to get there, it was unknown if there would be hazardous material surrounding Pluto – another, smaller moon, or a dense debris field, perhaps. But New Horizons succeeded brilliantly, and in doing so, truly revolutionized the way we see Pluto. The images that returned were the first high-resolution, color images of the world at the edge of our solar system, providing puzzles, answers, and more. Pluto went from a fuzzy blob to a fully developed little world nearly overnight (see color plate 9).

While we had known that Pluto had some color variation, no one could have predicted the terrain that New Horizons showed us. There was the heart-shaped bright plain, oddly missing any sign of cratering, the dark regions immediately next to it, valleys, and mountains.

A lack of craters meant that the surface must be young. *This* surface has erased all memory of impacts with other objects – which have almost certainly happened in the past – somehow. This may mean that Pluto is geologically active – somehow. That far away from the Sun, there shouldn't be enough heat to warm the surface of the dwarf planet. It's possible Pluto could have retained some heat from a high-energy collision, but we would have thought that 1) those impacts would have happened a long time ago and therefore 2) the heat should have already vanished into outer space. Pluto being the way it is means that one of these two assumptions is wrong, or there's another, third option we haven't thought up yet.

9. Just fifteen minutes after its closest approach to Pluto on July 14, 2015, NASA's New Horizons spacecraft looked back towards the Sun and captured a near-sunset view of the rugged, icy mountains and flat ice plains extending to Pluto's horizon. The smooth expanse of the informally named Sputnik Planum (right) is flanked to the west (left) by rugged mountains up to 11,000 feet (3,500 meters) high, including the informally named Norgay Montes in the foreground and Hillary Montes on the skyline. The image was taken from a distance of 11,000 miles (18,000 kilometers) to Pluto; the scene is 235 miles (380 kilometers) across. (NASA/Johns Hopkins University Applied Physics Laboratory/Southwest Research Institute)

In addition to the oddly flat plains, Pluto has mountains. Some of these are higher than 11,000 feet, and given how high they are, how steep, and the materials around on Pluto's surface to work with, they must be made of water ice. No other ices would remain so mountainous – they would have slumped with time. There's an abundance of ices to work with: for instance, nitrogen, which is rarely an ice on Earth (even in low-temperature chemistry experiments it's used as a coolant in its liquid form) is present as an ice all over the surface of Pluto. It seems that the nitrogen ices act on Pluto as glaciers do on Earth, flowing slowly along the surface. Pluto showed us an incredibly diverse and unexpectedly alive surface via New Horizons.

Pluto's atmosphere was another surprise – very little was known about it, as trying to observe it from Earth is extremely hard. The atmosphere we observed with New Horizons was not only much bigger than we thought (it extends much further from the surface than expected) but it has bizarre stripes in it. That's not an artifact of the image – those are truly in the atmosphere. It seems that some of the surface ice evaporates during daytime, and suspends itself in the atmosphere, where it lingers for a long enough time to build up these layers. If any gas escaped immediately to outer space, as we had thought it should, then Pluto's atmosphere should be tiny, a thin shell of gas sheathing a mountainous, glacier-y, young world. But we now know different.

Exoplanets: our sisters from other stars

If we wander outside our immediate family into the rest of our Milky Way galaxy, we begin to realize how many planetary cousins we have – planets which are circling other stars. To give them their proper name, they are extra-solar planets, or exoplanets. These planets have been difficult to detect for a long time, requiring either very sensitive measurements of the pull of (usually) a giant planet on the star it circles, or by watching for regularly occurring drops in the amount of light coming from a star, which tells us that something keeps passing in front of it.

These exoplanet-hunting efforts have recently been pushed into overdrive with the launch of the Kepler satellite in 2009. Kepler's primary mission was to watch for incredibly faint, regular drops in the brightness of more than 150,000 stars in the Milky Way. It has found 2,337 confirmed exoplanets, and more than 4,400 planet candidates, as of 2017. Kepler hunts in a very small area of the sky, so there are many more planetary cousins yet to be discovered. This marks a turning point in what has been a long, arduous hunt for these planetary relatives of ours

– the first tentative discoveries, later confirmed to be real exoplanets, were in the late 1980s.

These early discoveries were slow, painstaking measurements, and had to focus on individual stars for long stretches of time, to be sure that a planet was the best explanation for whatever signal you saw in the light. Kepler has allowed us to monitor the stars in a much more efficient way, since it can watch for flickers in the starlight from many more stars at once. The number of known planets outside our solar system has accordingly skyrocketed.

There are two main methods to find an exoplanet. The first is trying to see if you can catch the gravitational signature of a planet around the star you're interested in. This is easiest to do if you have a particularly massive planet circling your star, because in that case, the planet and the star are *both* orbiting their mutual center of mass. The center of mass can be thought of as the average position of all the atoms involved in the gravitational system. If the star is on its own, then its center of mass is precisely at its own center. If you add a planet into the mixture, then the positions of all the atoms which make up the planet will start to pull the average position away from the center of the star, which still holds the majority of the atoms. If you have a large enough planet, then the center of mass will be pulled far enough away from the center of the star that the star no longer appears to spin in place, but rotates wildly, like a set of keys spun on your finger. We can spot this wobbling motion because it causes very specific features within the light produced by that star to move along with its own motion. But in order to capture the motion of the star this way, you have to be consistently watching the star with an instrument capable of detecting the small motions of the features buried within the star's light. Because the wobble of the star gets more extreme with the most massive planets, this method tends to find that type of planet more than Earth-mass planetary cousins.

The other method for finding exoplanets is to wait for the planet to pass in front of the star, blocking out some of the star's light. You need a very sensitive camera, because the changes you're interested in are fractions of a percent dimmings as the planet crosses the star. With enough time, you can watch for regularly scheduled dimming of the same star, which flags up that you might have spotted a planet. This method finds *large* planets most easily. Large doesn't necessarily mean the most massive – a large fluffy planet works just as well as a large massive planet, but in general they do tend to track each other relatively well. To do this properly, you need to be taking regular measurements of individual stars over a long period of time, so that you can catch the repeated flickering of the star's light as it's blocked. This is exactly what the Kepler satellite has been doing. Kepler has one of the most fantastically precise cameras ever produced, able to detect changes to the brightness of a star's light to much greater detail than ever possible from the Earth, and its camera precision in combination with the efficiency with which it can observe the sky is what led to our exoplanet explosion.

The omnipresence of planets in the single patch of sky we've searched with Kepler has led some scientists to estimate that on average, every star has a planet. Some stars do not have planets, and other stars seem to have six, but on average, a planet per star seems to be a good starting point. Since the Milky Way is forming an average of about three new stars every year, which is not too bad for a galaxy of its size, we should be continually gaining a few planetary cousins somewhere within our galactic neighborhood.

How to grow life on a planet

With so many planets out there, the opportunities for life, even within our own galaxy, seems almost endless. But getting the set of conditions to come together which have been required for life to flourish on Earth is a considerably more challenging task than just having enough planets to work with.

On Earth, the biggest requirement for life is water. Our planet is very good at growing things in every possible location, so long as it's near or in liquid water. Life arose extremely quickly after the formation of the Earth (only about 200 million years after the impact that formed the Moon), which seems to indicate that once the Earth had a surface with liquid water on it, there were not a lot of other stumbling blocks to overcome before life could spring forth.

Liquid water itself is a challenge. It usually means that the planet must be in a relatively narrow distance window away from its star, and have a surface upon which the water can rest. Effectively, we need rocky planets at exactly the right distance from their sun, such that all the water doesn't freeze solid or evaporate away. Outside of that distance band in a solar system which allows for liquid water (called the habitable zone), there are precious few opportunities for liquid water to exist.★

....

★ That's not to say that there's no water elsewhere in our solar system – it's just not present in a *liquid* form. There is a class of asteroids which are also laden with water ice, behaving as an intermediate between the metallic asteroids and the primarily-ice comets. We have also found water ice in a number of permanently shadowed craters around our solar system. The permanent shade guarantees that the temperatures never rise high enough to evaporate the water to outer space. Any of these water-laden objects, if they were to come barreling into a planet or a growing proto-planet, could deliver water even if little had been present before.

There are, of course, a few exceptions, like Enceladus, a moon of Saturn, and Europa, a moon of Jupiter. Both of these moons are thought to have an ocean of liquid water under their icy surfaces, maintained entirely without help from the Sun. These small moons can maintain liquid water because the tidal forces from the massive planets they orbit are constantly stretching the rock at the cores of the moons. This stretching heats up the rock, and that heating provides the energy required to maintain liquid water. As a result, planetary scientists are very excited about the possibility of life on Europa and Enceladus, but in order to go check, we'll have to send a craft to those moons to (very carefully) go and look directly.

For our exoplanet cousins, this requirement of liquid water means that of all the thousands of planets that have been discovered, we're most interested in the rocky planets which sit in this magic distance window from their star where liquid water can exist. These planets are extremely hard to detect, and push the boundaries of the sensitivity of our telescopes. It seems, from the Kepler data that's been studied so far, that about 20% of all stars like our Sun have a rocky planet near enough to the star to have liquid water.

Proving that liquid water does in fact exist on those planets is more difficult still – you have to detect the signature of water in the atmosphere of a planet that is light years away. Proving the existence of *life* will be an even more difficult task, but once we begin to find lots of planets with liquid water on their surfaces, the odds are pretty good that one of them will contain life of some form. It will of course be much easier to search for life within our own solar system, since we can actually go to these places and see what's there directly.

THOUGHT EXPERIMENT:
Life gets weird sometimes

We often think about life beginning to develop in shallow seas of warm water, because that's the story of life on our parent planet, but to get a sense of how many other places life might be found on our exoplanetary cousins and planetary siblings, we should probably explore our own Earth's oddest and least-hospitable corners first. Many of them turn out to be inhabited, from the nearly frozen waters in Antarctica, to caustic caves, to the hot mud of Yellowstone.

Even our atmosphere isn't off-limits. We recently discovered that our own atmosphere seems to be rather full of living things, suspended surprisingly far up above us. A hurricane scouting plane took samples of the air a little over 6 miles (10 kilometers) above the surface, and found a phenomenal density of bacteria and fungi apparently thriving up there! At the very least, the bacteria they found were not all dead, which is a good start.

This was a surprise, because the higher up you go in the atmosphere, the less protection you have from high-energy ultraviolet (UV) radiation from our Sun. UV radiation is generally dangerous to life, because it is high enough in energy to ionize atoms and molecules, meaning it kicks electrons* out of their otherwise stable orbits; this can damage cells, causing them to mutate or die, depending on the severity of the damage. In humans this can lead to skin cells reproducing much faster than they should – it's one of the triggers for skin cancer. The atmosphere does a fairly good job of blocking most UV light, but the further up you go, the less protection there is. 6.8 miles above the surface, 75% of the mass of the atmosphere is below you, so this really is a very extreme, unprotected place for bacteria to survive. To find a large volume of bacteria, alive, seemingly unaffected by the UV dosage 6 miles up really was unexpected. At the moment, we think that Earth's storms are responsible for flinging so many bacteria nearly to the stratosphere, but it's the tiny mass of the bacteria which allows them to stay suspended up there.

---

* Electrons are a fundamental part of an atom, and are negatively charged. The flow of electrons is responsible for electricity.

Finding not-dead bacteria in our own atmosphere means that it's not entirely unreasonable to suggest that the same thing might happen in other atmospheres. Suspicions immediately turn to Venus, everyone's favorite 860°F, runaway greenhouse, volcano-ridden, battery-acid raining planet. Now that description, while accurate, does not paint a picture of a particularly habitable place. We don't, in fact, expect to find life at the surface, where we have lost each and every one of our probes to a combination of crushing and melting after only a couple of hours.

However, if you stay away from the surface, there's a layer in Venus' extraordinarily dense clouds which is positively balmy in temperature. It sits about 40 miles above the surface with a pressure about equal to that at the surface of the Earth, and is about standard room temperature. Unfortunately for humans, this is also the part of the atmosphere of Venus which rains sulfuric acid. This toxic acid rain evaporates before it hits the surface, leaving a catastrophic layer in the atmosphere where no humans would dare to pass.

For bacteria, however, this may not spell immediate demise, because our planet has a set of bacteria which thrive in the most unexpected places, and one class of them are fine with sulfuric acid. These particular bacteria live in caves, form stringy mats, eat sulfur compounds, and produce sulfuric acid as a byproduct.

They hang from the ceiling of the caves and are called snottites, or, if you prefer, snoticles. These caves are seriously unhealthy places for humans, both because of the general lack of oxygen and because of the sulfuric acid dripping from the ceiling (generally explorers need to be wearing heavy-duty protective gear and gas masks). But if a similar class of bacteria were present (ignoring how they got there) in the reasonable-temperature, reasonable-pressure cloud layer of Venus, they might be able to survive fairly well without worrying too much about the omnipresent sulfuric acid.

This sort of thinking isn't limited to Venus – it's just that it's the closest planet, we have the most information on it, and it's probably the easiest to explore. Jupiter has also been subjected to the same thought experiments. There's a long line of science fiction authors musing on the idea of Jovians, many of them exploring the idea of creatures living in the clouds. Though it's *extremely* unlikely that there are airborne jellyfish or cloud whales on Jupiter, it's more possible that if life is hanging around there, it might be microscopic life, suspended in the more charitable cloud layers. If we ever do find signatures of life hiding out in the cloud layers of our planetary siblings, that will be a strong hint that our exoplanet cousins might be hosts to similar forms of life.

So a planet's atmosphere may turn out to be one reasonable location for life, but there's one more place we might find it – in stars which missed the amount of mass you need in order to start fusion burning in their cores: brown dwarfs (see page 156). The coldest of them are really quite cold: the most extreme surface temperature is somewhere between –54°F and 9°F (–48°C to –12°C), which is right about the limit of the coldest survivable conditions for extreme-loving bacteria on Earth. Not all brown dwarfs would be suitable; they would need to be as Jupiter-like as possible, which happens only with the smallest of them, where the boundary between a Jupiter-like planet and a failed star is the fuzziest. But given what we know about stellar atmospheres today, if life could thrive in the high atmosphere of gas giants like Jupiter, then the lowest-mass stars, which may yet outnumber stars like our own, could be the home of star-borne life.

Of course, all of this is purely a thought experiment until we can go exploring and see for ourselves. There are missions that have been designed with present technology to go look for life in Venus' clouds. There are others to go check the oceans of Europa. We may find that life among our Earth's siblings is not so rare as we had thought.

Observational problems

The planet-hunting satellite Kepler has been searching a very small patch of our galaxy, and even its finely honed systems can only detect Earth-like planets out to 3,000 light years* distant.

Our Milky Way galaxy is about 50,000 light years from center to edge (so about 100,000 light years across), and the next nearest large galaxy is Andromeda, sitting about 2.5 million light years away from us. While we expect to find a planet around pretty much every star in our galaxy, we haven't been able to survey even a sizable fraction of the Milky Way, let alone the stars in Andromeda, which would be exponentially more difficult to observe. The furthest solid detection of an exoplanet through any means is still only about 21,000 light years away, well within our own galaxy.

However, the limitations of the speed of light mean that any images we get of exoplanets are just as out of date as they are distant from us. A planet that we see at 10,000 light years distance from us will be an image that has traveled for 10,000 years.

---

* A light year is the distance that light, traveling at about 670 million miles an hour, can traverse in one year. One light year is therefore 5.9 trillion miles, or 9.5 trillion kilometers.

On a geological timescale, 10,000 years is just a blip of time – the Earth was pretty much in the same shape as it is now, though we humans had made fewer changes to its surface. On a *human* timescale, 10,000 years has made a big difference. 10,000 years puts us back into the Neolithic era – the end of the Stone Age, around the time when pottery was developing, and we were beginning to cultivate plants for agriculture. So an intelligent civilization, 10,000 light years distant, that is just now looking for other life in the universe would spy our Earth as a rocky planet with an atmosphere, far enough away from the Sun that water could exist in our atmosphere, and if they managed to examine that atmosphere, they would notice that it is mostly nitrogen, with some oxygen and carbon dioxide in it as well, and that it contains water vapor. They would not be able to tell that there are creatures on that planet that are 10,000 years away from developing the internet, neurosurgery, and machines able to detect tiny distortions in space itself.

This kind of time delay is one of the reasons that scientists get extra excited when they find a nearby rocky planet that might be able to have liquid water on its surface – if the planet is close to us, then the time delay isn't as bad as a more distant planet. (Though, if we're being honest, their excitement is primarily because it is *much* easier to observe these nearby planets in any degree of detail – the farther away you are from the Earth, the harder these measurements get.) We only managed to detect the contents of the atmosphere of a slightly-bigger-than-Earth planet for the first time in February of 2016. Unfortunately that planet is totally devoid of water, having an atmosphere of mostly hydrogen and helium, with some hydrogen cyanide thrown in for extra poisonous flavor. This planet is only 60 light years away, so that image of it is only out of date as far as 1958. This particular planet won't have evolved into a friendlier, life-hosting planet in such a short time.

The further out you go, the worse this time delay gets. In practical terms, we're unlikely to be able to do anything, be it communicate with, visit, or otherwise signal a very distant civilization, if we did detect some signs of life on a distant exoplanet. The speed of light is the very best we can do – if we were to beam a radio message to a distant world, that radio wave would speed outwards from the Earth at precisely the speed of light, and no faster (because radio waves are a form of light). The speed of light is a fundamental speed limit within our universe – no information travels faster than the rate at which light could traverse the distance.

And the same is true of any spacefaring civilization many billions of light years away from us. If a super-intelligent civilization out there could build an impossibly large telescope, and had the power (and the time) to detect planets orbiting stars in a distant galaxy, and they happened to point it at our Earth, they might see a very different world than the one we currently inhabit. If they happened to be 2.5 billion light years distant, our planet's atmosphere would be in the middle of a dramatic change. This was when the Earth was undergoing the 'oxygen catastrophe' – the earliest photosynthetic bacteria were dumping oxygen into the atmosphere faster than it could be absorbed, and oxygen was slowly building up. As oxygen was a toxic byproduct to the single-celled life which had been living in a delightfully oxygen-free environment, they would have had to adapt or die off. Observations of our planet from that distance would be able only to tell the observer that our planet existed, it had water in its atmosphere, and that we journeyed around our star every 365 days. There wouldn't be so much as a hint of our space-exploring future.

Rules of engagement

The unlikeliness of encountering life on one of our exoplanetary cousins hasn't kept us from setting up some guidelines for what to do if we *did* find signs of life on another world that isn't the Earth. These guidelines aren't binding in an international court of law. There *is* a branch of law for dealing with outer space, called International Space Law, which is a set of rules that the United Nations has laid out and all UN member countries have agreed to abide by. It is a string of very sensible rules for behavior in space, like 'do not put nuclear weapons in space', 'your country cannot claim a piece of space', 'you cannot build a military base on the Moon', and 'space is for all countries'.

Most of the guidelines for what to do in case we find or are found by alien life have a similar set of instructions, whether it be found in the SETI Institution's version, entitled 'Protocols for an ETI Signal Detection' (SETI standing for Search for ExtraTerrestrial Intelligence), or the 'Declaration of Principles Concerning Activities Following the Detection of Extraterrestrial Intelligence', which has been signed off on by a whole slew of administrative bodies. But really, they've only signed off on agreeing to be guided by the guidelines.

The guidelines go roughly like this: if you think you have found a signal from extraterrestrial life, please make extra, seriously, super-duper sure that what you think is an ET signal doesn't have any other more plausible explanation. The discovery of the star KIC 8462852,★ which was

---

★ Also known as Boyajian's star, after the author of the paper who first described this phenomenon.

found to be extremely inconsistent in the amount of light from it that was reaching Earth, triggered a huge wave of excitement, as it was rapidly dubbed the 'alien megastructure star', after the suggestion was made that perhaps the light from the star was being blocked by a large, artificial structure of extraterrestrial, intelligent origin. With further observations, that star is now thought to have a very thick disk of extremely fine dust orbiting it (which can block large fractions of the star's light), possibly in combination with a swarm of comets, which can account for the faster flickering that was seen. Even with a relatively rapid scientific turnaround away from the alien construction project, there were a huge number of articles on the possibility. I can only imagine the sort of media explosion and cultural whiplash we would have on our hands if we announced that we had discovered intelligent life. This is a 'be careful' initial guideline.

Secondly, pass what you think to be a detection to other scientists, and let them check your work. This checking of your work may mean getting more data, more observations, or just someone else verifying that they can reproduce your numbers. At this stage, please do not leak this to the media. It may well be that what you have found is very interesting, but not aliens. If a scientist immediately announced, 'I found aliens', and then it turned out that what they had actually found was something astrophysically bizarre, but not aliens, there would be an instant loss of trust in that person. This second stage is to make sure that at the very least, reliable numbers and information are announced.

Next: if the signal has not gone away and still looks like alien life after some serious checking, you should tell the Secretary General of the UN, and a whole string of scientific bodies, which will let the rest of the scientific community know. You should also swiftly tell the rest of the world. But hey – if it was your data that did the discovering, you get the honor of hosting a very high-profile press conference, and will probably spend the next year and a half (at *least*) answering questions.

Publish your data. However you detected intelligent aliens, please tell other scientists so we can do more of it. This means explaining what you did at conferences and probably making your data public. Also, please put your data in as many places as possible. The last thing we would want at this stage is to lose it. Put it in ten places. Put it everywhere. Please do not lose the evidence of aliens to a hard drive failure.

If the aliens were detected on a particular frequency (say we heard a radio transmission of theirs), that frequency should probably be protected so we can keep listening without too much interference. Radio quiet areas are pretty rare nowadays, and a lot of things can cause radio waves (a bizarre chirp detected by a radio-sensitive telescope was recently proven to be generated by someone opening a microwave door to stop it).

And last in the guidelines, you do not get to write back to the aliens. Not without getting the majority of the planet on board, anyhow. At that point, contacting aliens has an impact on the whole human race, not just science, and not just the country that the scientist who discovered it lives in. The decision of what to do next should be discussed by everyone.

But because these are only guidelines, if some private company happened to do the discovering, and disregarded all of the above rules, the way things sit currently there would be no legal ramifications. The guidelines are also unlikely to be needed for many centuries.

This unlikeliness simply comes from the fact that our galaxy is a very, very large place. Our planet is supremely unremarkable in terms of the star we orbit, our location in the galaxy, and our galaxy's location, so it seems likely a similar configuration should have produced itself elsewhere. The trouble lies in the vastness of how many 'elsewhere' options there are.

Astronomers traditionally tackle the unlikeliness of finding alien life by means of the Drake Equation, which takes all the pieces we think need to align for life to evolve, and multiplies together all the probabilities that they'll occur in the same place at the same time within our galaxy.

In a simple form, it asks the following series of questions:

- From the number of stars in our galaxy, what fraction of those stars will have planets?
- On average, how many planets are at just the right distance from their star, where liquid water can exist at the surface?
- Of the planets with liquid water hanging around, how many of those should we expect to have any form of life, no matter how simple?
- What fraction of those planets with life will also have intelligent life?
- And what fraction of those intelligent life forms are still around now and potentially able to communicate with us?

There are at least 100 billion stars in our galaxy, and hundreds of billions of galaxies in the universe. As noted earlier, virtually every star we've looked at in our galaxy has at least one planet, which means we're dealing with an overwhelmingly large number of planets in the universe. Our Earth went from 'liquid surface water' to 'life' in about 125 million years, a cosmic blink of an eye – that doesn't seem to be the hard part. The tricky part is getting from bacteria to a species that contemplates the skies above. But even if you're extremely pessimistic about the fraction of planets that have any form of life, and pessimistic about the fraction of planets with intelligent life, the sheer number of planets out there dominates these calculations. There simply has to be other life out there, and if not in our galaxy, in another.

But the Milky Way is *huge*. As we saw above, it takes light 50,000 years to travel from the center to the edge of the galaxy: 100,000 years to make it all the way across. And the distances between galaxies are even more extreme. Light from the Andromeda galaxy takes 2.5 million years to reach us, and that's our nearest neighbor. 2.5 million years ago, humans were only at *Homo habilis*, mastering early stone tools. We don't have any record of modern humans from earlier than 200,000 years ago. Our first radio telescope was built all of 80 years ago.

We can do a few more rough calculations to figure out how bad the problem is. Let's assume that you work out that by optimistic numbers, there should be fifteen intelligent, communicable civilizations just in our galaxy. If those civilizations are randomly scattered around the galaxy, they're separated by about 23,000 light years. If there's only one civilization per galaxy, you're back to separations of millions of light years. Communication between civilizations (even in the most optimistic of cases, where we send a signal out at the speed of light) would be impossible.

*

Regardless of the unlikelihood of encountering civilized life in our galaxy, the fact that our solar system is accompanied around the Milky Way by a swarm of planetary cousins, each orbiting their own star, means that our cosmic family is large indeed. Each of those cousins owes their existence to the formation of their own star, much as we owe our existence to the cloud of gas which collapsed to form our Sun. But how exactly do these stars work, and how different from our own star can they be? This requires some study of the next tier of our family tree, and a trip further back into our genealogy.

4

STARS

The shining creators of the elements

For all the endless fascination of our neighboring planets, it's the stars that make planets possible, so let's take a trip one step up our family tree and learn a bit about the stars themselves.

Stars are tremendous collections of gas, compressed into such a high density that the temperatures at their cores allow for the construction of new elements in their centers. Without the stars, life on our planet would certainly not exist, as the sunlight that warms our planet enough for liquid water is the after-effect of the Sun's own construction of these new elements – and the Sun is of course a star.

If we humans are the children of the Earth, the Sun is our grandparent. Our planets would not have formed except around a star. The leftover remnants of the gas cloud which collapsed to form our Sun went into forming the planets, as we saw in the last chapter. Without the initial collapse of the gas to form the Sun, there would have been no dense gas and dust ready to form the Earth and its sibling planets.

In the way of many grandparents, the Sun fed our planet and its life-form grandchildren – in our case, a rich diet of sunlight for a watery world, allowing photosynthetic plant life, and more complex animal life in turn, to flourish.

The Sun: Sol

So where does the sunlight emanating from the Earth's parent star come from? To answer that we're going to have to take a bit of a dive into the center of the star, through a number of layers of stellar material, and into the core, where it all begins.

Our Sun, known formally as Sol, is a pretty good prototype for how stars are constructed and how they behave; it is a middle-aged star, in the middle of the longest period of stability it will ever have, and it is not particularly large or small, when we compare it to all the rest of the stars in our galactic neighborhood. As a result, its interior is also pretty average, and a journey into its core will give us some insight into how most stars work.

Stars are mostly made of hydrogen gas, largely out of sheer convenience – hydrogen is the most abundant gas in the universe. If you collapse any random patch of gas in the universe, you're likely to have gotten yourself a big pile of hydrogen, perhaps with a few other elements tossed into the mix in small quantities.

Our star is no exception; it's mostly hydrogen plasma. Plasma is a phase of matter one step more extreme than a gas, where the electrons which normally make up an atom have gained so much energy that they've been able to vault away from their nucleus, and are wandering freely. The nucleus of the hydrogen atom, now that its electron has left, remains as a simple proton.*

As we journey to the core of the star, this hydrogen is subject to hotter and hotter temperatures, and higher and higher pressures. A star is a very precise balancing act between the crushing force of gravity, pulling all particles inwards, and a resistant pressure that comes from both heat and a reluctance of particles to be pressed too close together, which forces electrons and protons further apart from each other.

In the very center of the star, the heat and pressure are sufficiently high that the hydrogen nuclei slam together at such high energies that they can stick together. This

---

* A proton is a component of an atom, found in the core of an element. The number of protons tells us what element the atom is. Protons are positively charged. The simplest atom, hydrogen, is a single proton and a single electron.

is nuclear fusion. It's a gradual process which involves the release of light and the conversion of protons into neutrons,* but at the end of the day, you find yourself with a fully constructed helium nucleus: two protons and two neutrons, all stuck together. This smashing together of hydrogen and helium nuclei can, in principle, be used to build more complex atoms, but the heat and pressure required at the center of the star in order to build those complex atoms is higher than we see in the Sun. For a star the size of our Sun, the building of hydrogen into helium is as complex as it gets.

This fusion process only occurs in the dense core of the Sun, which extends from the very center out about one fifth of the way to the surface; from edge to edge this places the core at a width of 172,915 miles (278,280 km). Large, but nothing when compared to the 864,576 miles that the Sun itself spans. Every single beam of light that the Sun produces comes from this small region in the very center of our grandparent star.

---

* Neutrons are another critical component of an atom, found in the core of the atom. They have no electric charge. Changing the number of neutrons in an atom changes the isotope of the atom, but does not change the element.

For a beam of light, escaping the core of the Sun is not a difficult task; this region is totally transparent to light. Two invincible friends sitting in the Sun's core would have no more trouble seeing each other than they would through thin air. It's the middle zone that serves as a roadblock for light: the radiation zone.

The radiation blockade

These zones are very uninventively named, and so the radiation zone is simply a zone surrounding the core of the Sun, where energy is moved through radiation. Radiation itself is something we usually think of in terms of cancer treatments or nuclear waste. Fundamentally, it involves light (photons) serving as carriers for energy; high-energy photons leave a very energetic area, and travel to a less energetic area, bumping into something and depositing their excess energy. This process then evens out the energy levels everywhere. Closer to home, this is the method behind a conventional oven, or a sunburn, or holding your hand over a hot surface to see if it's safe to touch. For a sunburn, a photon of light has left the Sun, traveled 93 million miles through space, through the atmosphere of the Earth, and bumped into your skin. Your skin absorbs the energy that the sunlight carried, and given enough time and enough absorptions, your skin will begin to protest at this energy dumping, which is damaging its cells. Living cells do not like radiation much, which is why we humans depend so thoroughly on the protection of our atmosphere to survive.

Light is caught by these same rules in this area of the Sun. Each photon moves from the hot core, bounces into some of the hot plasma in the radiation zone, and gets absorbed. What we have is a stream of newly formed light coming from the fusion reactions happening in the core, ready to flow out of the Sun, which then hits the boundary of the radiation zone, only to be absorbed by the material there. But it's hot here, near the core of the Sun, and so the energy is quickly spat back out in the form of a photon, going in a new random direction. That photon doesn't have far to go before it runs into a new plasma particle, and gets absorbed again. And it's still hot here, and it gets spat back out again, but now in another random direction. Any photon we could choose to follow will get trapped in this cycle of absorption followed by emission in a random direction.

Because the material here is very dense, the photons never have far to travel before their next bounce, and each time they hit something new, they're redirected in a random direction. This directionless motion is called a random

10. A random walk with 2,500 steps. (László Németh, Creative Commons CC0 1.0 Universal Public Domain Dedication)

walk, and if you have a goal in mind, it is one of the least effective ways of getting there (see image above).

For a photon of light, trying to escape the Sun and perhaps eventually land on a planet, the radiation zone is a trap that lasts several hundred thousand years.

At last, a roiling rise to the surface

But let's say 200,000 years have passed and our lucky photon has escaped the radiation zone. I did say that there were three zones. The third is the convection zone. (I did also say these were uninventively named.)

Convection, like radiation, is a method of moving energy around, but instead of doing everything through a photon pack mule, convection moves the particles around themselves. You don't need the plasma of the Sun to touch your skin to get a radiation burn – the photons do that for you – but if you want to mix hot and cold substances together, you're going to need to shuffle around the molecules. The most common everyday example of this is to look at a pot of boiling water, where the fluid in the pot is roiled and moving around with the bubbles rising to the surface (not just sitting still), and donating its energy to its surroundings.

Convection takes over in the Sun when we've reached a sufficient distance from the core that everything has cooled down a bit, and so photons of light leaving the radiation zone at long last are absorbed by the plasma in the convection zone, and that material itself is heated by that absorption. The plasma begins to rise, like a bubble of water in our boiling pot, pushing directly towards the surface of the Sun. You can even see these bubbles on the surface of the Sun, if you have a very powerful solar telescope.

As the hot material rises, the plasma closer to the surface of the Sun, which is cooler, sinks back down to replace it. The bubbling up of hot material only takes about a week, which, in comparison to the hundred thousand years it took for the light to escape the radiation zone, is positively instantaneous. Once the bubbles reach their peak, the light stored inside this heated material is free to leave the Sun entirely and stream outwards into space in whatever direction that tiny piece of the star's surface happens to be pointing. This point, where the Sun becomes transparent to light again, is what we've termed the 'photosphere', and it's this surface that we see in images of the Sun. There is a tenuous, thin atmosphere beyond the photosphere, made of very hot gases (much like the rest of the Sun), but the best time to see that atmosphere is during a solar eclipse, when the bulk of the Sun's light is being intercepted by the Moon.

Our magnetic grandparent

All of this so far is a good description of the path that light takes to make it from creation to escaping, but it's not yet the complete story of our cosmic grandparent. What's missing is the magnetic field of the Sun.

The magnetic field of the Earth is a bit more familiar to us, and we discussed it in more detail in Chapter 1. The Earth's magnetic field orients our compasses, and protects our planet from radiation which would be harmful to life, producing the aurora effect as some of these high-energy particles from space end up slamming into our atmosphere.

The Earth's magnetic field is pretty straightforward – it has a North and a South Pole like a bar magnet (see image below), and so the protective bubble that surrounds our Earth is orderly and neat. It's like an over-inflated donut that encompasses a wide swath of space around our planet. The Sun's magnetic field is not nearly so neat.

There's a critical difference between the magnets we can stick to our refrigerators and the magnetic nature of a planet or a star, and it all comes down to the fact that fridge magnets are solid, and planetary cores are not. For fridge magnets, the magnetism is effectively frozen in place within the atoms of that magnet. This means that the magnetic field surrounding that magnet is totally static – it doesn't change over time, and no property of the magnet will change its strength or shape.

The magnetic fields of the Earth and the Sun are very different beasts. On Earth, the magnetic field is formed by the motions of magnetic nickel and iron in the core of the planet, which generates an electric current. Because it's

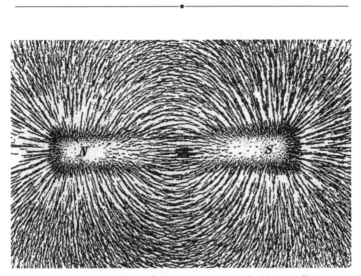

11. The magnetic field of a bar magnet revealed by iron filings on paper. A sheet of paper is laid on top of a bar magnet and iron filings are sprinkled on it. The needle-shaped filings align with their long axis parallel to the magnetic field. They clump together in long strings, showing the direction of the magnetic field lines at each point. (Newton Henry Black)

generated by the motion of material deep within the Earth, the magnetic field is not nearly so static as a fridge magnet. Magnetic North, which is the location that compasses point to, therefore drifts by a few degrees every year, and wanders around the general area of Geographic North.★

★ Geographic North is the physical point where no stars would rise or set, since you're exactly at the top of the spinning Earth. Magnetic North is the top end of the magnetic field, and is most often not at the same place.

On the Sun, the magnetic field is also generated by the motions of material within its interior which produce an electric current. So you might think that the Sun's magnetism should behave like the Earth's magnetic field. But there's a critical difference between the Earth's behavior and the Sun's; a generational divide, if you like. The Earth, very conveniently for us humans, has a solid surface. The solid surface means that the planet has to rotate as a coherent unit. There's no way for the Bahamas to have a shorter rotation time than England – both islands are fixed to the rock below, and both have a 24-hour day.

On the Sun, there's no rock to anchor the rotation speed of the star to a single value. As the entire contents of the Sun are a superheated plasma, it is relatively easy for something at the equator of the Sun to move with respect to something at the poles.

In fact, this is what we see. The equator of the Sun rotates a few days faster than the poles do. Unlike the Earth, whose surface is mostly non-magnetic, the Sun's surface is welded to its magnetic field, and so the rotation differences between poles and equator mean that the magnetic field gets pulled along into an increasingly twisted configuration. After about a month, when the Sun has done one full rotation, instead of having magnetic field lines which smoothly travel from pole to pole in the same kind of donut shape as the Earth, our star has twisted the middle of its magnetic field considerably to the side.

If you let the Sun continue to rotate (not that we have much choice in what the Sun does), these magnetic field lines will get more and more twisted over to one side, until they no longer resemble any kind of ordered pole-to-pole path. At this point, the tangled magnetic field is unstable enough that it will begin to pop little sub-loops of magnetism out of the surface of the Sun. The more repetitions of this month-long twisting process that go by, the more of these little loops there are, bursting out of the surface. These sub-loops are what give our solar grandparent its spots. Where the loops touch the surface of the Sun is where we see dark patches, known as sunspots.

Sunspots are slightly cooler regions on the surface of the star. The magnetic field loops produce a pressure outwards at the surface, and while normally the Sun is a consistent density, a consistent temperature, and a consistent pressure over its entire surface, when a magnetic field loop appears, that even spread of properties is slightly changed. Once the gas at the surface is pressed outwards slightly, the gas in that region can cool down. And because of their lower temperature, sunspots appear darker than the surrounding area. As the Sun's magnetic field becomes increasingly tangled, more and more of these dark spots appear.

Stellar outbursts

Sunspots are not static objects on the surface of the Sun, and as the magnetic field of the star changes, their size and shape can change in response. The Sun's magnetic field has another trick to play as well: those sub-loops of magnetism aren't permanent.

If the magnetic field of the Sun is sufficiently tangled, these loops pushing out of the surface can themselves become twisted. If the loop crosses itself, the magnetic field will snap itself into a new, shorter configuration – where the loop connected to itself will be the new, shorter magnetic pathway. If the loop has crossed near the base, the magnetic field will create a very low, flat loop, just above the surface of the Sun, where previously there had been a more lengthy, extended loop. Above that, we have a newly freed strand.

This free strand is made of protons, electrons, and other bits of the surface of the Sun, which up until the magnetic reconnect, had been able to flow smoothly along the magnetic loop. Without the loop's presence, the strand is no longer attached to the surface of the Sun. Nor will it stay hovering over the surface of the Sun for very long; there's a lot of energy released at the moment that the magnetic field reconnects into a flatter loop, and that energy goes into violently kicking this strand outward, into the space surrounding the Sun.

If this reconnection event is large enough, the resulting blowout of the disconnected strand of the Sun is called a coronal mass ejection. Any strong burst often goes by the name 'solar storm'. Most of the time, the Earth is well out of the way of such outbursts, and so the particles swing outwards through the solar system without encountering much resistance.

If, however, the alignment of events is unlucky, sometimes the Earth winds up directly in the path of one of these high-speed particle dumps. The Earth's protective magnetic field takes the brunt of these blasts, so we humans on the surface of the planet are usually in no danger. Our orbiting electronics, on the other hand, are much more exposed to these kinds of phenomena.

We have put a lot of satellites in orbit around our planet, and every one is a very delicate set of electronics. If the Sun sends one of these blasts of material towards our satellites, suddenly the electronics within them are being bombarded with far more charge than they were designed to withstand.

The Sun sends a certain amount of protons and electrons our way every day, but the density involved in a coronal mass ejection is so high that satellites, if unprotected, can short-circuit. The constant bombardment of electrons means that any metal parts of the satellite can build up a charge on the satellite itself, like shuffling in your socks on carpet. If enough charge builds up, the satellite will short itself out, destroying critical components. This is mostly an issue for satellites in very high orbits (GPS satellites are one example), and that are less deeply embedded in the magnetic field of the Earth.

The International Space Station, for instance, is in a low enough orbit that it's mostly protected from these kinds of solar storms by the magnetic field, but every so often, there's a strong enough storm that the astronauts on the ISS take shelter in more strongly shielded portions of the structure, just for good measure.

Solar storms and coronal mass ejections are almost never strong enough to press back against the magnetic field of the Earth with such intensity that electronics on the ground are affected, but it *has* happened. In 1989, there was a stupendously strong solar storm, which managed to affect the magnetic field of the Earth to such an extent that power cables started to carry current in a way that they weren't designed for.

A series of systems then failed to manage this unexpected current, and the failures led to 6 million people losing power in Quebec. In repairing the damage done in this event, the system has been redesigned so that it is harder for space weather to break it so easily, but the modern power grid is widely considered unprepared for another, similar event. This is one of many reasons to maintain a fleet of telescopes which monitor our star – if we can see these bursts when they leave the Sun, we'll have a few days of warning before they reach the orbit of the Earth.

These solar storms tend to concentrate in a few-year period when the Sun's magnetic field is at its most tangled. But there's always an end in sight – after the eleventh year of tangling up its magnetic field, the Sun resets itself. The magnetic field turns over and begins again as a smooth magnetic loop, from the top pole to the bottom pole. Once the magnetic field resets, the Sun goes quiet and sunspot-free for a few years while it gradually restarts the process of tangling itself up again. After another eleven years, there's another reset and turn-over, completing the cycle.

The view that we have on our Sun, with its sunspots, solar storms, coronal mass ejections, and complex magnetic field maps, is a unique one. No other star in the universe has been studied in as much detail. But these local features aren't the end of it – in fact the Sun influences the space surrounding our entire solar system. A steady stream of particles blows outwards from the Sun, forming what we know as the solar wind. This wind marks the interior of a bubble which encases our home, and all of our planetary siblings, bounded by the outer limits of the Sun's magnetic field.

A larger perspective on our Sun

Within the bubble where our solar system sits, space is entirely ruled by the Sun's influence. The combination of the Sun's magnetic field and the solar wind means that inside the solar system, we don't feel the influence of the other stars in our galaxy. Here in the solar system, space is relatively calm, aside from the occasional blast from the surface of our Sun. We're sitting inside our grandparent's house with the heat on, watching a storm rage outside our windows.

This boundary between our Sun's regime and the rest of the galaxy is one way to define the edge of the solar system. We've given it a name: the heliosphere (from the Greek meaning 'sun-sphere'). Beyond it, the direction of the cosmic wind changes, signaling a shift into what's called 'interstellar space' – the space between sibling stars.

We're still learning about the nature of the edge of our solar system. We've only aimed a few spacecraft out that far, and to date, only Voyager 1 has crossed into interstellar space (see color plate 10). Voyager 1 was determined to have officially made it out of a transition region and reached the other side in August 2012. Voyager 2, its duplicate craft, is on a slightly different pathway out of the solar system and has not yet exited the solar system completely, even though they were only launched sixteen days apart from each other.

We once thought that Voyager 1 would leave the solar system all at once, very much like stepping through the boundary of a soap bubble. Instead, Voyager encountered a much more gradual transition, where a few of the expected observations for leaving the solar system were seen, but not all of them at once. There was, however, a particularly dramatic moment on August 25, 2012, where the amount of plasma the spacecraft was plowing through increased sharply, as did the number of cosmic rays.

As we've seen, cosmic rays are very high-energy particles, which typically arrive into the solar system from elsewhere in the Milky Way galaxy. Our star shields us from many of these cosmic rays, so once Voyager had left the Sun's protective bubble, it made sense that it would have encountered many more. For Voyager, this transition occurred at around 121 astronomical units (i.e. 121 times farther away than the Sun is from the Earth), a distance that takes light itself just over sixteen and a half hours to traverse. Compared to the vast distances to the nearest stars, which it takes light entire years to cross, even this large stellar bubble shrinks to a very local space.

With no more craft on the way to the edge of the solar system, other than New Horizons, and with a journey time of more than ten years to get there, should we decide to go exploring the boundaries of our solar bubble once more, there will be a wait. Nonetheless, there's still useful information coming from Voyager 1. Voyager 2 should be crossing into interstellar space in the next few years, hopefully with a few more functional and active instruments on board. With more instruments, Voyager 2 will be able to tell us even more about this boundary to our solar system.

Once we've reached the outer edges of our solar system, we've exhausted the region of space that our own Sun can influence, and it's time to turn to its siblings, its companions in the orbital dance around our galaxy, and the objects which illuminate our dark nights – the other stars.

Our Sun and its siblings

Our particular star is typical of its siblings, but what that *doesn't* mean is that all the stars are similar. An average human is of average height and weight, but humans are incredibly varied, beyond that average set of values. The stars in the night sky are equally diverse, ranging from stars tens of times more massive than our Sun, which glow a toxic, cancer-inducing, blue color, to the faint red dwarfs, less than a tenth the size of the Sun, which can glow a reddish purple, not even as warm as the surface of Venus in some cases.

Our family of stars is collected into our galaxy, the

Milky Way. Each of our stellar family members, along with any planets they have in their tow, orbits the center of the galaxy on its own path. Even with the huge numbers of stars in the Milky Way, the sheer volume of space in the galaxy means that no two stars will ever hit each other. Our galaxy is shaped in an almost two-dimensional disk – proportionally speaking, the Milky Way is much thinner than a sheet of paper. The family of stars that inhabits this disk was made possible by the galaxy's immense gravitational pull, which collected gas to itself in the early universe. Without that initial pulling together, it would be difficult to find the concentrations of gas that you need to construct a star.

Most of the stars in our galaxy formed here, either in the Milky Way as we know it, or at a much earlier time, before the galaxy was fully constructed. But it is this common pool of gas and dust that brings these stars together as family.

If we want to differentiate a stellar sibling from the rest of its family, there must be a critical set of information to know about that star. The simplest one is color.

A star's color tells you many things

Stars produce light of many colors; our own star produces every color in the rainbow – very literally. But unless we go to some effort to split all these colors of light apart from each other, either through a scientific instrument or through the remarkably effective prism of the raindrop, the light which reaches our eyeballs from our Sun appears white, because all of these colors of light have mixed together, and the average color is pretty much smack in the middle of the visible spectrum.

Some stars in the night sky are very obviously not white-light stars – Betelgeuse, one of the bright stars in Orion's shoulder, is a red supergiant with an orange glow even to the unaided eye. The mixture of light coming from this star has much less blue than our grandparent Sol. The lack of blue light being produced means that the red light can tilt the average color over towards a more reddish-orange.

We can dig a little deeper than this – rather than just looking at the average color coming from the star, we can break apart the light into its respective colors, and check exactly how *much* more red light is coming from a star, relative to blue. For a different star, we might measure just how much more *blue* light is coming to us, relative to red. From this kind of measurement we can determine the most commonly produced color of light for each star. A star which produces a lot of red light, like Betelgeuse, will have its most commonly produced color sitting very close to what we would call red, or perhaps slightly beyond what we can see, in the infrared.

A blue star, on the other hand, will have a most common color which is either blue or beyond blue, in the ultraviolet. And our white-light Sol? Sol's most common color is an eye-bending yellow-green. Rather than using the *average* color to discuss stars, it's usually this *most frequent* color that is used. This is how our star became known as a yellow star, even though the light is white to our eyes.

The color of a star isn't randomly assigned – it's a direct result of a few of its other properties. The color of each of the stellar siblings in our galaxy is determined directly by the temperature at the surface of the star. Temperature is in turn ruled by the mass of the star, along with how aged it is.

The tie between temperature and color is one that we encounter here on Earth as well, and it wanders in our language. We have turns of phrase for things which are 'red-hot' and 'white-hot', and we know that metal glowing orange is hotter than something glowing red, though both are dangerous to human skin. What we may be less familiar with is how far this relationship goes. Red-hot, for a star, is as cool as they come; the plasma that makes up the star is hot enough to glow of its own accord, but not hot enough to create a lot of blue light. Blue light needs more energy to produce, so at low temperatures, which means relatively low energies, there's just not enough energy to pump out a lot of blue light.

As the temperature of the star rises, the amount of available energy rises as well, and so more and more high-energy (blue) light is able to be produced. Stars will brighten from a deep red, to a brilliant white, to a vivid blue, if they are hot enough. The hottest stars produce a large amount of ultraviolet light, beyond the visible range, in addition to bathing their surroundings with blue light. These bright blue stars are relatively hostile to life; the moderate dose of UV light we receive from our Sun is already dangerous to humans, and luckily our atmosphere's ozone layer interferes with this light, blocking the higher energies from reaching the surface. With too much high-energy radiation, however, atmospheres can be destroyed, and life forms can be irradiated into oblivion. While these stars might have their own planetary children, hot blue stars are unlikely to have life-form grandchildren unless that life is particularly well sheltered or much more resilient against radiation than we are.

So color is the direct consequence of temperature, and temperature is, in turn, dictated by the mass of the star. For the set of our stellar family members which are, like our Sun, in a stable middle phase of their lifetimes, this is a fairly direct relationship. The more massive these stars are, the hotter their surface.

A star that contains a lot of mass is a stronger gravitational influence on the space surrounding it than a lower-mass star. This increased influence is also present on the material of the star itself. As we saw earlier, all stars are a balancing act between the inward, crushing pressure of gravity and the outward pressure of the stuff of the star resisting that inward force. With a stronger inward force of gravity, because there's more mass in the star, in order for the star to be stable, you must counterbalance that with an increased outward pressure, or the star will collapse and become spatially smaller, and denser.

Any time you compress a gas (and all stars are made of gas), the temperature rises, and as the temperature rises, the outward pressure increases. So if the star collapses down a little bit, the temperature inside the star increases, which increases the pressure pushing back against the force of gravity. As long as this happens slowly, the star can find a balancing point where inward gravity and outward pressure cancel each other out.

The spatial size of this end result also matters – a massive star ends up taking up more physical space once it finds its balancing point. If you have a larger glowing object, you illuminate the space surrounding your object more strongly.

This tie between the size of a star and its brightness is so strong that if you doubled the size of a star, you would expect it to become four times as bright, if the temperature remained the same. There's a much stronger relationship between temperature and brightness: if you double the temperature, but keep the size the same, you'd expect the brightness to increase by sixteen times. Both temperature and size change smoothly with each other, though, so you wouldn't expect to be able to keep one of them the same as you dial up the other.

A stellar description

With the combination of the temperature of the star, as we trace it by its color, and the brightness of the star, which traces its physical size, we can piece together a lot of information about the vast family of stars that inhabit the Milky Way.

Stars that contain more mass than the Sun, and are physically larger, have a much higher gravitational pull than our own star. In resistance to the greater inward force, the outward pressure has grown much higher, cranking the temperature up internally. The surface temperature of such a star mirrors that internal temperature, making it glow a hotter, bluer, hue than our own Sun. Because the star is also physically larger, its brightness is also more intense than the Sun.

On the other hand, we have Sol's smaller siblings, stars with much less mass than our star. These have less mass to crush inwards, and so they find a balance with less outward pressure from the material within the star. This less-dense balance means that the core of the star is never pressed into extremely high temperatures. With a lower temperature in the core, the surface is even cooler, and only manages to glow a dim red. The smaller amount of material also means a smaller star, which drops the *brightness*, on top of it only glowing a duller red color.

These two scenarios, plus our Sun, give a range of stellar possibilities: from bright, large, and blue, to faint, small, and red. But is that the only set of stars you can get? Or are there exceptions to this? To answer that question we have to take a census of the nearby family of stars, and check what the rest of them are doing. If we find that the brightest stars are *all* blue, and the faintest of them are *all* red, then the description above is as far as we need to go.

A family census like this is best illustrated by charting the brightness of a large number of stars against their respective colors. This kind of chart has been named the Hertzsprung-Russell diagram, or H-R diagram for short. If our faint red to bright blue relationship describes all stars, then we should expect to see a diagonal line running across the diagram, with nothing anywhere else.

We do see the diagonal line (color plate 12). This line is what we call the 'main sequence' of stars, and it is where all of the stars which are stably burning hydrogen in their cores sit. The vast majority of the stellar family members in our neighborhood and galaxy sit along that line. However, our Sun has a few unusual siblings, which are found in a totally different part of this diagram.

It's useful to categorize stars more carefully than just into red, blue, and yellow like our Sun, and the H-R diagram is a good place to start dividing the stellar family into sets of cousins, where the stars within that group are all pretty similar to each other and can be compared as a group to other groups.

The first broad division we can make is a horizontal one, across the middle of the H-R diagram. The top half of the diagram is where all the extremely bright, physically large stars in our galaxy can be placed. The bottom half gives us the smaller, dimmer stars. These smaller stars are formally given the term dwarf stars, distinguished from their larger cousins, called giants. This giant/dwarf divide is particularly useful for stars which glow an intense red, as a red star can either be a small star along the main sequence, or one of the giant stars, well away from this main sequence line. If we use the color of the star, and whether it is a giant or a dwarf, we can easily describe any star in our home galaxy – and beyond.

Our Sun, which sits roughly in the middle of the diagram, is just on the dwarf side of the giant/dwarf divide; technically speaking, our star is a yellow dwarf. We've talked extensively about how our own star works, and its twin stars, other yellow dwarfs (they bear tangled names, from Tau Ceti to HD 147513), so far seem to be operating in the same way.

But what of the other types of star? Let's begin with the other dwarfs. Once we exit the yellow dwarf area, we have a few stellar colors which remain: red dwarfs, brown dwarfs, and the only small stellar sibling of ours which is bluer – white dwarfs.

Red dwarfs

Red dwarfs are one of the closest relatives of our Sun – just a little less massive than Sol, they also sit along the main sequence, towards the fainter, redder end. Red dwarfs, much like the Sun, burn hydrogen at their cores, but have gathered much less material to themselves in their formation. This smaller amount of mass means that the pressures and temperatures at their cores are lower, and they burn at a slower rate as a result. The slower burn means that these dim, redder stars will outlive their bluer siblings. Their smaller mass means that they're relatively easy to form, and are expected to be among the most numerous type of star in our stellar family. It's hard to count them directly, because they are so faint and red, but we know that our nearest stellar neighbor, Proxima Centauri, is a red dwarf star.

THOUGHT EXPERIMENT:
Divide the Sun in half

Imagine we split our Sun in half. We could end up with two largish red dwarf stars orbiting each other at the center of our solar system. But as we've mentioned, the relationship between the amount of light that's produced and the mass of the star is not a one-to-one trade-off. If I cut a star in half, I halve its mass, and I drop the amount of light produced by that star by 90%. The drop in temperature between Sol and a red dwarf is significant: about 40% less. Our Sun's surface temperature is 9,980°F, or 5,527°C, whereas a red dwarf of half the mass only checks in at 6,200°F or 3,426°C. The brightness of a star is *very* sensitive to temperature, as we mentioned earlier, and our red dwarfs are also smaller than our Sun.

A double red dwarf system, though gravitationally similar to our own Sun, would only be able to produce

20% of the light that the Sun generates. That is a pretty dramatic reduction in the brightness of the center of our solar system, which would, in turn, dramatically shift the places surrounding that double-star system where liquid water could exist. A planet like Earth, orbiting that double-star system at a distance of 93 million miles, would not be able to keep water from freezing. As far as we can tell, liquid water is the primary requirement for life to exist, so an Earthlike planet around such stellar parental twins would be frozen solid.

As we saw in Chapter 3, the liquid water zone is calculated as the shell of space surrounding a star where there's enough energy reaching the surface of the planet for water to melt, but not so much that the water would evaporate away as steam and be lost to space. The formal name for this region of space is the habitable zone, and it's the focus for many searches for plausible Earthlike cousins outside our solar system.

In a double red dwarf star system, the habitable zone would be so close to the stars that if we placed our own planets around those twin stars, *Mercury* would be in the liquid water zone. Mercury's orbit sits between 0.3 and 0.46 times the distance between the Earth and the Sun. The habitable zone would range between almost these same numbers: 0.32 au, if the planet was getting light from only one star, up to 0.44 au, if it was getting sunlight from both of them. So Earth would be completely frozen, but the water ice trapped in Mercury's darkest craters would be able to thaw, forming crater lakes – and given enough time, possibly life.

Brown dwarfs

If you keep pushing down the main sequence to even fainter and dimmer stars, you arrive at an interesting corner of our stellar family. By definition, stars are objects which can construct hydrogen into more elaborate elements through nuclear fusion, and this process operates more efficiently the hotter and denser the core of your star happens to be. So what happens if you let the internal temperature of the star drop, and drop?

At some point, the core of the star just isn't hot and dense enough to build anything. Brown dwarfs are, in essence, failed stars. They collapsed, gathered material to themselves, and began to heat up, but there just wasn't enough material around to build a fully-fledged star, to be able to begin building up the elemental complexity of our universe.

Unable to generate their own heat, brown dwarfs have to be content with simply radiating out into the cosmos what heat they generated during their collapse, slowly cooling to the average temperature of the universe. This process takes an extraordinarily long time, so we don't expect any of them have succeeded in getting rid of *all* their heat – but if they had, we'd call them our Sun's black dwarf siblings.

Brown dwarfs are essentially very large Jupiters, though brown dwarfs are counted as stars while their Jupiter counterparts are always planets. They can be 'cold' even on human scales: a brown dwarf found in 2011 was only as warm as a cup of coffee. Their masses are so much smaller than our own star that instead of being discussed in units of multiples or fractions of the mass of the Sun, we use the mass of Jupiter instead. If the star is more than 83 times the mass of Jupiter (or 0.08 times the mass of the Sun), it can start to burn hydrogen in its core. Brown dwarfs then range from this upper limit down to somewhere around 13 times the size of Jupiter.

There's nothing special about this 13 Jupiter mass cut-off for how small a brown dwarf can get. At that size, it's debatable whether you should call this collection of gas a very large planet (which would usually orbit a fully-fledged star), or a very small star (which would usually be at the center of its solar system). If the brown dwarf/large Jupiter happens to be orbiting a much larger star which *is* burning hydrogen in its core, it's even more ambiguous. Binary stars are common in our galaxy – like twins, being born out of the same cloud of gas, so it's not out of the question to have a brown dwarf twinned with a yellow dwarf like our Sun. But when the brown dwarf is particularly small, like a large planet, orbits like a large planet, and isn't generating its own heat (like a planet), it's easy to begin to classify these objects as planets instead of stars.

The smallest brown dwarfs, if they should be considered planets, would be the children of the stars; but if they are the smallest stars, then they must be placed one tier up, as the potential parents to their own planetary children. This kind of bridge across the generational gap between stars and planets is an unusual one, and it's part of what makes brown dwarfs so intriguing. The largest planets in our solar system (Jupiter, Saturn, Uranus, and Neptune) are typically dubbed 'gas giants', and it's in the gap between Jupiter-like planets and the stars which illuminate those planets that brown dwarfs find themselves.

White dwarfs

Aside from the red giants, white dwarfs make up the other population of stars which are very far from the main sequence – in fact they're the only kind of star which falls *below* the main sequence.

This is a weird place for a star to be, because it means that the star is, for some reason, burning relatively hot but is also extremely faint. If the star is hot, the only way for it to be so faint is for it to be very, very small. As we've seen, small stars usually don't have a lot of mass, which is how you get faint red stars – so why are these burning so hot?

White dwarfs, unlike our Sun, are not burning hydrogen in their cores. In fact, they're the leftover *core* of a star, hyper-compressed into an extra dense state. White dwarfs are often about the physical size of the Earth, but contain large fractions of the mass of the Sun. This ultra-high density means that the material within the star is existing in one of the most extreme places in the universe. It also gives us the explanation for the location on the diagram. The white dwarf siblings of our Sun are *tiny*, which explains their dimness in our universe. However, their huge amount of mass accounts for their white-hot glow, given the intense inward crushing force that gravity is exerting on the star.

A normal star has an outward pressure to balance gravity, as we have seen, and white dwarfs must have this as well, or they would simply continue to get smaller as time progresses. We see them as stable siblings of our Sun, so what generates this resistant force? The mere presence of electrons is what's doing it. At the incredible densities of a white dwarf, electrons resist further gravitational crushing. If you wanted to compress a white dwarf further down, you'd need to compress electrons closer in to each other.

The typical white dwarf doesn't have the gravitational force required to mash electrons any closer together, and so they can put up an effective resistance, holding the star from becoming any smaller.

Know them by name

It's one thing to be able to describe our stellar family in groups of similarly behaved stars, and another thing entirely to give them each a name. As much as we call our own star 'the Sun', its formal designation is 'Sol'. We might know a few other names of Sol's siblings – the North Star, for instance, is named Polaris, and Betelgeuse, a red supergiant star, sits at the constellation Orion's shoulder, as we noted earlier.

But what about the other stars? If you continue digging for names, you'll start to find ones that are less familiar to your tongue, unless you happen to speak Arabic. We don't have to leave the constellation of Orion to find a few: the stars in Orion's belt are named Alnitak, Alnilam, and Mintaka. These names are old; we began to collect the names of stars into catalogs as early as Ptolemy, in 150 CE. Ptolemy's catalog was translated from Greek into Arabic, and that Arabic catalog was then circulated around western Europe.

Many of the Arabic names are simply descriptions of the location of the star within the constellation – stating that they are in the belt, for instance. In the process of coming from the Arabic translation back into western Europe, many of the spellings were altered, and so some of these oldest names either have multiple spellings, or have converged to a spelling which has only a loose resemblance to the original Arabic. Mintaka and Alnitak, for instance, could have been more accurately written out as al-Mantaqa and an-Nitāq.

But these are only the brightest of the night-sky stars, and of course our telescopes have improved immensely since the days of Ptolemy, vastly growing the list of our known stellar family.

An attempt to make a comprehensive list of all the stars in the night sky was undertaken in the 1600s by Johann Bayer, and this has since taken on his name as the Bayer Designation system. The idea was to adopt a more rigorous approach, where for every constellation, each star could be named in a more catalog-friendly way, with a Greek letter to mark its brightness and a name to mark its constellation. The brightest stars in Orion then became Alpha Orionis, Beta Orionis, and so on, and fainter stars were assigned Greek letters further down the alphabet. However, this alphabetical assignment is not always exactly in line with the brightness of the stars. Any stars which vary in brightness can be particularly troublesome, if you are trying to sort them in order.

The Bayer Designation was used for a number of years, but we have begun to take a huge number of photographs of the night sky, at ever fainter depths, and any attempt to maintain an all-sky catalog of stars with a consistent naming scheme has essentially been abandoned. Most of the stars that we identify in the night sky are now 'named' with an alphanumeric identifier, which marks the exact survey of the sky which discovered them, and often their location. These stars will not gain more pronounceable names unless something very unusual about them is later discovered, and the scientists investigating that abnormality need a faster way of referring to the star.

Constellations

The constellations themselves, even though we've spent many hundreds of years basing our stellar family's naming schemes on them, are nothing more than the storytelling human brain at work on the night sky. With a couple of exceptions, the stars in the constellations are no more related to each other than the lights of a plane flying at night are to the stars.

The stars in our constellations are usually fairly bright, which means that they can't be on opposite sides of the galaxy from each other, and they remain related by the simple fact that they are all in the same galaxy. But not being on opposite sides of the galaxy and being really and truly lined up in the night sky are two very different things.

The constellations are simply temporary alignments of the stars in our neighborhood, and they only exist as we see them from our particular vantage point on the galaxy. If our solar system were even a little bit differently placed in the galaxy, those constellations would transform into a completely different jumble of stars relatively rapidly.

Orion is a good example of this. It's a bright constellation, with a number of the brighter stars in the night sky of the northern hemisphere contained within it. In winter it's easily recognized, even in cities where the light pollution can be severe. And yet its stars are *hundreds* of light years apart from each other.

Betelgeuse, the bright red star in Orion's shoulder, is about 640 light years from Earth. In contrast, Bellatrix, the other shoulder star, is only 200 light years from Earth, three times closer. Mintaka, the rightmost star in the belt, is a whopping 1,200 light years away, twice as far as Betelgeuse. If you put those distances into a model of the constellation, and rotate around it with an arbitrary camera, you can see how dramatically the constellation changes as your perspective changes.★

This isn't unique to Orion. Effectively any constellation you could point to would do the same thing as we moved around it. The slightest shift in where our solar system is placed in the galaxy would scramble the entire constellation. We're not ever going to move the solar system, but there's another way for the constellations to scramble themselves: waiting for a long, long time.

The Big Dipper (also known as the Plough) is also made up of stars which are entirely unrelated to each other, but those stars are *all* much closer to us than the stars in Orion. The seven brightest stars that make up the Dipper range from 58 light years to 124 light years distant from Earth. Much in the same way as trees by the side of the road appear to zip past your eyes faster than objects in the distance, stars which are closer to our position appear to move much faster in the skies than distant stars. So the stars in the Big Dipper will appear to move over time more rapidly than those in Orion. We can measure the speeds at which these stellar family members are moving, and predict where they will be in the future. In another 100,000 years, the handle on the Dipper will be bent

★ https://www.youtube.com/watch?v=lD-5ZOipE48

under, and the front of the Dipper bent outwards. It's a relatively quick destruction of an astronomically temporary alignment.

Globular clusters

Many of the stars in the night sky are solitary, completing their orbits around the center of our galaxy on their own, or perhaps with a few planetary children to keep them company. Our Sun is one such star – we have no secondary sun in our skies, and the Sun orbits the center of the galaxy with the solar system coming along with it. But not all stars are so isolated, and there's another assembly of stars that's worth a look before we explore all the varied ways that stars can come to the ends of their lifetimes.

Globular clusters are very dense, very old clusters of stars, where all the stars are gravitationally bound to each other, so that it will be very difficult for them to ever escape. These are a family group; gravitationally tied together, and usually made up of stars that are all the same age, it's thought that they formed nearly all at once and have remained clustered together as a tight-knit community. These clusters are often found outside the main disk of the galaxy, in a spherical arrangement instead of the flat, disky arrangement of the rest of the stars in the Milky Way (see color plate 13). They can be quite isolated from the rest of the stellar family within the galaxy, and within them, a much older population of stars exists, many billions of years predating our own Sun.

Globular clusters are home to a few hundred thousand stars on average, and our Milky Way hosts a few hundred of them. (The galaxy Andromeda, our closest galactic point of comparison, has at least 60 globular clusters, though we're bound to miss some when we're hunting for a faint collection of stars 2.5 million light years away.) Sometime in the formation of the Milky Way, these clusters of stars were formed and trapped into spherical groups around the galaxy.

It must have been early in the formation of the galaxy, because the stars that populate globular clusters are very, very old. We mentioned earlier that small, red stars take much longer to exhaust their supply of hydrogen gas, and can stick around as stars for a very long time. Some of the stars in globular clusters appear to date back nearly 13 billion years – about as old as you can get. In fact, for a time, the age of these stars was used to age the universe as a whole. If you've got stars hanging around that are 13 billion years old, the universe had to be older than that. But their exact formation is still a mystery, and how they manage to remain unperturbed for such a long stretch of time, while the universe around them has changed so drastically, is another point of study.

One thing is for sure: if the globular cluster was going to collapse onto itself, or flatten itself, or change in any way over time, it's had more than enough time to do so. Since we're seeing them *now* as they are, they must be stable in their current setup. So what does keep all these densely crowded stars from collapsing inwards onto themselves?

We've talked a lot so far about the influence of gravity on individual stars, so if you crowd 100,000 stars into a fairly small region of space, you might very reasonably expect that gravity should pull them all down into each other, collapsing the whole lot into a messy superstar at the very center of what was once a globular cluster. If gravity were the only force at work here, that's exactly what would happen.

But gravity is almost never the only force at work, and globular clusters are no exception. In this case, the missing piece of the puzzle is that every single one of the 100,000+ stars in the globular cluster is in motion. And with motion, we can defeat gravity in the same way that the International Space Station stays aloft: by orbiting. If you have enough motion 'sideways' from the direction that gravity is trying to pull you, the end result is an oval orbit around the center of mass. For the ISS, this is a very nearly circular orbit around our parent planet. For a star in a globular cluster, its oval pathway takes it in loops around the center of the cluster. As long as that sideways motion persists, the star won't fall to the center of the group.

Every single one of the stellar cousins in the globular cluster is using the exact same strategy, with absolutely no rhyme or reason to the direction of 'sideways' they choose to go. The randomness of the orbits of all the stars is what makes the cluster look so spherical. If the stars all had the same definition of 'sideways', like the planets in our solar system did, you'd wind up with a disk, and the cluster would look flattened.

This very specific kind of structure – a random set of orbits, from a large number of objects, creating a fuzzy, spherical haze of stars, is given the technical name of a 'pressure supported' structure. There's no *real* pressure here, like there is on the inside of each of the stars, but it serves as a distinction from the ordered rotation of a solar system disk or galaxy disk, and it is still a resistance to the force of gravity. (Astronomers are bad at naming things.)

So it is the motion of the stars that lets these globular clusters remain so stable over the lifetime of the universe. But globular clusters aren't totally immune to changes over the course of 13 billion years.

The stars within a globular cluster are more tightly

packed by far than the stars within the disk of the galaxy – typically they wander within a light year of each other, though at the very center, where the stars are the densest, they can come as close as the distance between the Sun and Pluto.* (This is not *very* close, but in astronomical terms it's snug.) As densely packed as they are, the stars do occasionally interact with each other when they get close enough; as they swing by each other, the gravitational force will tug on them both. If the two stars are about the same size and mass, but one has a much faster orbit around the center of the cluster, it may overtake the slower star, if they're moving in the same general direction. The gravitational tug between the two will slow down the faster star, donate that energy to the slower star, and speed it up. Because they have similar mass, the two stars can end up traveling much closer to the same speed after their encounter.

But what happens if the second star is much less massive? In that case, the same amount of energy donated to the small star can bounce that slower star to a much higher speed. The change in speed of any star is related to the energy donated, and its mass. The less energy, the less of a boost; and the less massive the star is, the more of a boost

---

* Pluto is 39.5 times further from the Sun than the distance between the Earth and the Sun. Or, if you prefer, about 3.7 billion miles.

it gets. The momentum of the smaller star is less able to resist changes in speed, and off it zooms. This is the same principle behind slingshotting spacecraft around Jupiter; we steal a small amount of energy from Jupiter, but because the craft is so little, it can increase its speed dramatically.

If you let this energy donation process proceed over enough time, what happens is that the massive stars are always donating energy to less massive stars, and gradually slowing down. Meanwhile, the less massive stars, which are on the receiving end of all these donations, gradually speed up. These speeds translate into changes in their respective orbits. If the star has slowed down, gravity has an upper hand, and pulls the star down into the depths of the cluster. If the star has sped up, it will end up flinging itself around the outskirts of the cluster at high speeds, in sharp contrast to the now lethargic orbits of the largest stars. The globular cluster has effectively sorted itself, with all the heaviest stars at its center, and all the lightest, smallest stars out at the edges – a process called 'mass segregation'.

There's one other thing that can cause a lot of change within a globular cluster. The clusters are all in orbit around a parent galaxy – and sometimes they wander too close. The orbits of globular clusters around their galaxy are extremely long, and many of them seem to have survived many such orbits, but it's not always a safe passage.

If the cluster travels too close to the galaxy, the galaxy's gravitational force can shear apart its outer layers. If the cluster has already sorted itself, that means it loses the lightest, smallest stars from its outer edge. These are pulled into a faint stream, barely visible even to powerful telescopes, and leaving the high-mass, larger stars at the nucleus of the cluster. At least four such globulars exist within our own Milky Way.

*

The stars in our night skies have always been stable features of human observations, but they certainly won't live for ever. Over the history of the universe, many generations of stars have lived, burned, and died, scattering material back to their home galaxies. But not all stars act the same when they reach the ends of their lives, so before we step up our family tree once more, let's look into the final stages of our stellar family.

5

STELLAR DEATHS

Explosions, balloonings, and fadings

The lifetimes of the stars dwarf our human ones: from the first fusion in its core to the end, our Sun will have provided around 8 billion years of relative stability. But all stars have a fixed lifespan, tied closely to their mass and the rate at which they run through the hydrogen available to them.

The main sequence of stars, that diagonal line from bright and blue to faint and red, is composed entirely of the siblings of our Sun which are stable and burning hydrogen into heavier elements in their cores. These stars maintain an outward pressure resisting the gravitational force pressing inward *because* they are burning so much hydrogen, and keeping the internal temperature high. As time goes on, more and more of the available hydrogen is consumed and turned primarily into helium. At some stage, there isn't enough hydrogen around to continue burning.

This marks the beginning of the end for a star. It must, at this point, undergo some change. Without the ability to continue burning hydrogen into helium in its core, a large part of the resistance to gravity is gone along with it. We should expect the star to compress itself as gravity wins out.

It does; the star's interior is crushed to even higher densities, pressing the plasma into a hotter, more confined space. This crushing will continue until a region surrounding the core of the star – now filled with helium – can begin to burn hydrogen. The core may stay inert, because as far as the hydrogen-burning process is concerned, helium is the ash left over after a campfire. If the star is large enough, the pressures in the center may increase enough so that the core can begin a brand-new burning process, building helium into elements like carbon and oxygen.

Our grandparent Sol will be able to burn helium like this, but the more massive siblings of the Sun will be able to push even further. Once the helium in the core of the star is exhausted, the star will collapse again, triggering hydrogen burning in a new shell of material even further from the core, plus helium burning in the shell where hydrogen had been burning, and the carbon and oxygen ash from the helium burning may ignite. This process of exhausting the fuel at the core, allowing the star to collapse further and igniting another layer, allows extremely massive stars (more than eight times the mass of the Sun) to build their way up to iron ash in their cores. That is the end of the line – the pressures and temperatures required to build the heavier elements cannot be produced through gravitational pressure alone.

Even with just two shells of burning, as our Sun will do, there's a *lot* of extra pressure coming outwards from within the star. A large amount of the mass of the star is in its depths, where the material is compressed. At the surface, there's less mass, and it's much more tenuous. With a boost to the amount of light being produced in the depths of the star, this outer layer is hit with an extra strong outward push. Not being very massive, and being relatively far from the majority of the mass of the star, these outer layers are forced outwards.

Several things are now happening at once. As the star's outer layers expand, they cool. It's the opposite effect that's happening in the core, which is heating up as it compresses. This takes the surface of the star from a white or blue color down to an orange or a red, as its surface temperature drops. On the other hand, the interior of the star is burning through helium and hydrogen, both at a furious pace, and the amount of light being produced has ramped up. The star ends up, even after cooling down its surface, thousands of times brighter than it was back when it was only burning hydrogen.

With all these changes going on, the star is going to move on our diagram (color plate 12). No longer behaving like its main sequence siblings, the end-of-life star moves to the right as it cools to a redder color, and up, to the brighter end, as its light-producing reactions grow more rapid. We've created a red giant star.

All stars, except the very smallest red dwarfs, will pass through this phase on their way out of stellar life. Our own star should do so in the next 4.5 billion years or so.

The end of our planet

The expansion of a star into a red giant is no small thing – when our star reaches this phase of its life, the surface of the Sun will extend 200 times further out than it currently does, roughly to the orbit of Mars. The Earth is not expected to survive the death of its parent star. Even if the Earth wanders a bit further from the center of the solar system, most predictions do not expect it to totally escape the atmosphere of the expanding Sun. If the Earth is caught in its outer layers, our star will devour the planet, vaporizing the rock we live on. Venus and Mercury, our planetary neighbors, have no chance of escaping disintegration by star.

This won't be an abrupt end of our planet – the Earth

will become gradually more and more inhospitable as billions of years run on. Even before our star finishes burning hydrogen, it will be slowly evolving. Since the Sun has begun building helium out of hydrogen in its core, it has increased its brightness by about 10% for every billion years it's spent burning hydrogen. Increased brightness means the amount of heat our planet receives is also greater. With enough heat, the water on the surface of our planet will begin to evaporate.

A 10% increase in brightness is enough to change the location of the habitable zone around our star, placing it outside the Earth's orbit. As soon as the Earth exits the habitable zone, we no longer have a planet which can stably store liquid water on its surface, and the evaporation of the oceans will begin in earnest. By the time the Sun stops burning hydrogen in its core and becomes a fully-fledged red giant, Mars will be in the habitable zone, and the Earth will be *much* too hot to maintain water on its surface.

As the oceans evaporate, the water will instead be held in the atmosphere. Water is an excellent greenhouse gas, much like carbon dioxide, and will serve to trap even more heat around the planet. With the heat trap in place, more and more water will evaporate to join the atmospheric vapor, increasing the greenhouse effect until the oceans are dry. The atmosphere will be a muggy, superheated, water-saturated mess. The water molecules that have made their way to the upper atmosphere will get bombarded with high-energy particles from the Sun, breaking them apart. Once the hydrogen and oxygen which make up the water molecules are freed, they can escape from the atmosphere into space. Over time, this bombardment of the atmosphere by the Sun will bleed the atmosphere dry of water.

The trigger point, when the amount of light the Sun sends to the surface of our planet becomes too great for stable water, is predicted to hit in about a billion years or so. The exact numbers depend on who you talk to. What differs between predictions is how fast this whole process unspools. Our planet will certainly walk this path, but some scientists suggest that the Earth will be roasted well before the 1 billion year mark, because the rocks, oceans, and plate tectonics may all speed the drying and heating of the planet. On the other hand, some forms of life have already proven themselves to be extremely durable and well able to maintain a foothold in odd corners of the Earth, living off of chemicals like the deep sea creatures of today. A billion years 'or thereabouts' is as good a baseline as any for the end time for our sibling life forms on our planet.

However much damage and destruction our evolving star will wreak on the neighborhood, our Sun is still only a medium- to small-sized star, and so its ballooning out into the solar system is still relatively small-scale destruction. To explore just how big a star can get, and just how many planets could be lost into the atmosphere of an expanding giant, we must turn our eyes to a much larger member of the stellar family: UY Scuti.

A giant among giants

UY Scuti is one of the very largest stars we know of. It's found in the constellation Scutum, near Sagittarius, along the line of the Milky Way in our night sky. UY Scuti isn't the most *massive* star − its gravitational clout is outdone by many others* − but it appears to be taking up the most space of any star in our repertoire. UY Scuti is not a light-weight: it's a good 30 times more massive than our own star. However, because its outermost layers are already being nudged outwards, the material which makes up the star already covers a region *1,700* times larger than the Sun.

Being nearly 2,000 times bigger than our own Sun places UY Scuti into a new class of star − a supergiant. Betelgeuse, the red star in Orion's shoulder, is one of the

....

* The Most Massive Star record is currently held by R136a1, which is measured at 265 times the mass of our own Sun.

most obvious supergiant stars in the night sky. Most super-giants are about half as large as UY Scuti, but that still places them more than 800 times larger than the Sun. The promotion of red giants to red supergiants is largely a nod to how much bigger they are than anything the Sun will ever become.

The size itself can be a tricky thing to measure, not least because these tremendously large siblings to the Sun also physically pulse, growing and contracting over time. This 1,700 number is a rough average size, and UY Scuti will exceed or shrink below this value every few years as it goes through its cycle of expansion and contraction.

But we can't understate the size of UY Scuti. From the center to the edge is equivalent to 750 million miles, or nearly eight times the distance between the Earth and the Sun. That size is large enough that the star would extend past Jupiter if it were at the center of our solar system.

I mentioned the 'edge' of the star, but this a much harder thing to define for a red giant or supergiant than it is for a star like our own. The surface of any star is defined as the point at which light can stream freely outwards into space. But for supergiants and red giants more generally, the star is gradually losing its hold on its outer layers, as they are pressed away and then become so distant that the gravity of the star's core has very little influence over them.

Red giants can lose a considerable amount of mass this way, and it means that many of these stars are surrounded by veils of their former surface, forming into their own personal nebula.* For a star which is the size of UY Scuti, this nebula of lost gas extends *much* further out. In our solar system, its cloud of gas would extend out 400 times further than the distance between the Earth and the Sun – ten times further out than even Pluto's distant orbit.

Stars the size of our own Sun won't create a gas cloud quite as large as UY Scuti's – Sol quite simply doesn't have the amount of gas required to spread itself so widely. Red giants can only maintain themselves while they have material to burn, whether that be helium, for stars like the Sun, or heavier elements for more massive stars. When they run out of material, they must find a new resting place. For stars which are less than eight times the mass of our Sun, the result of the next gravitational collapse is the formation of a white dwarf, in what used to be the core of the star (see page 158). The rest of the star, both the diffuse atmosphere which has already been loosed from it and the remainder of the star's shells of gas, will be flung outward into the glowing, symmetrical clouds of gas we call planetary nebulae.

---

* A nebula is a cloud of gas, illuminated by the presence of nearby stars.

Smaller stars, our own Sun included, all meet their ends this way.

Supernovae

We saw above how the members of our stellar family which are on the small side – somewhere less than eight times the mass of our star – will find their end by the last-minute puffing out of a planetary nebula, leaving the hot core of a white dwarf behind.

But what of the most massive stars? Ones like UY Scuti, or Betelgeuse, which are much more massive than our Sun?

Once these stars reach the end of their fuel, and can burn nothing more, they have a different fate in store for them. The most massive stars produce some of the most energetic events in the universe – supernovae. A supernova can allow a single star to outshine the entire galaxy's worth of its sibling stars, several billion in number. Supernovae are fascinating events, in part because we can watch them unfold on short timescales by our human standards, after the star itself has existed for billions of years in its hydrogen-burning phase. Supernovae grow in brightness over a period of about a week – a phase called the 'rising time'. The star will glow its brightest for a few days, and then drop back into dimness over another couple of weeks.

What's happened to the star that makes it glow so brightly? Let's go back to the center of our red giant, which in massive stars is burning silicon into nickel. The form of nickel produced in this reaction is unstable, and decays into iron. Once the star runs out of silicon to burn in its core, gravity will again take over, crushing the core down once more. The star can't push back – there's nothing else it can do with the nickel and iron that's built up. This is the same process that generates a white dwarf, but there's a lot more material involved here. If the very core of the star is larger than 1.4 times the mass of our Sun, this white dwarf-esque configuration is unstable, and gravity can compress this central core more violently and into a denser object. The collapse from something about as dense as a white dwarf to something much more dense happens rapidly, within a couple of seconds. The material in a shell around the core suddenly lurches, and with nothing underneath it, it falls rapidly inwards to reconnect with the smaller core of the star.

The rapid compression of this shell of material heats it up to a tremendous temperature very quickly, and it releases a huge burst of energy.

Meanwhile, the core of the star stops collapsing, and the free fall of the gas shell is abruptly stopped as well. Coming in with such speed, much of the material bounces away again, in a ricochet outwards. Having gathered so much energy, this material is now reflected back squarely at the outer parts of the red giant. It's a one-way path – the sheer energy contained in the rebounding gas blows the star apart, leaving the collapsed core on its own.

What we typically see as the supernova remnant in these images (color plate 14) is the shock front of the ejected material hitting any gas and dust which surrounded that star. As the surrounding gas and dust will be a little differently arranged for each star, the shock front will have a slightly different shape for each supernova as well (color plate 11). Depending on the density of the material the shock front is running into, the expansion of the front will continue at different speeds. Inside that edge, there can be a reverse shock, where material has raced up behind the shock front, and bounced off of that back again towards the core of the star which remained behind. This second rebound can heat up the material on the inside of the supernova bubble to very high temperatures.

The remains of the outer layers of the star, once ejected, are lost to that star. Most of it is simply too far away and moving too fast to collapse back onto the remnant of the core of the star.

The gas that gets flung out of the star and into its neighborhood does so at a very rapid speed after getting kicked

13. The object shown in this beautiful Hubble image, dubbed Messier 54, could be just another globular cluster, but this dense and faint group of stars was in fact the first globular cluster found outside our galaxy. Discovered by the famous astronomer Charles Messier in 1778, Messier 54 belongs to a satellite of the Milky Way called the Sagittarius Dwarf Elliptical Galaxy. (ESA/Hubble, NASA)

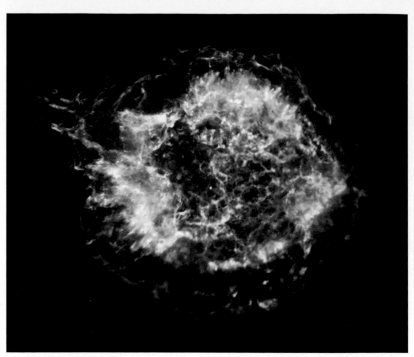

14. Cassiopeia A: A supernova remnant located about 10,000 light years from Earth. (NASA/CXC/SAO)

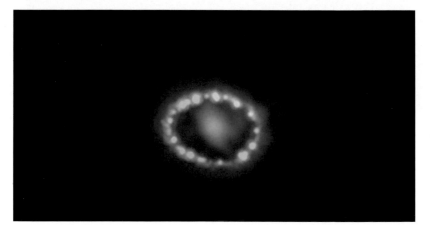

15. This image shows the remnant of Supernova 1987A seen in light of very different wavelengths. ALMA data (in red) shows newly formed dust in the centre of the remnant. Hubble (in green) and Chandra (in blue) data show the expanding shockwave. (ALMA (ESO/NAOJ/NRAO)/A. Angelich; visible light image: NASA/ESA Hubble Space Telescope; X-ray image: NASA Chandra X-Ray Observatory; CC BY 4.0

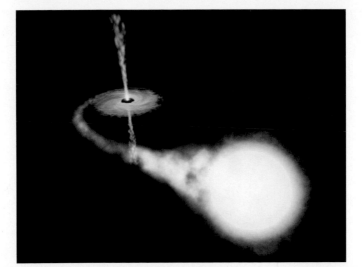

16. An artist's rendition of a black hole with an orbiting companion star. Mass from the companion star is drawn towards the black hole, forming an accretion disk. (ESA, NASA, and Felix Mirabel (French Atomic Energy Commission and Institute for Astronomy and Space Physics/Conicet of Argentina))

17. The large Whirlpool Galaxy (left) is known for its sharply defined spiral arms. Their prominence could be the result of the Whirlpool's gravitational tug-of-war with its smaller companion galaxy (right). (NASA, ESA, S. Beckwith (STScI), The Hubble Heritage Team (STScI/AURA))

18. This Hubble Space Telescope image of the face-on spiral galaxy Messier 101 (M101) is the largest and most detailed photo of a spiral galaxy that has ever been released from Hubble. (NASA, ESA, K. Kuntz (JHU), F. Bresolin (University of Hawai'i), J. Trauger (Jet Propulsion Lab), J. Mould (NOAO), Y.-H. Chu (University of Illinois, Urbana), STScI; Canada–France–Hawaii Telescope/J.-C. Cuillandre/Coelum; G. Jacoby, B. Bohannan, M. Hanna/NOAO/AURA/NSF)

19. This artist's conception illustrates one of the most primitive supermassive black holes known (central black dot) at the core of a young, star-rich galaxy. Astronomers using NASA's Spitzer Space Telescope have uncovered two of these early objects, dating back to about 13 billion years ago. (NASA/JPL-Caltech)

20. The Hubble Ultra Deep Field is an image of a small region of space in the constellation Fornax, composited from Hubble Space Telescope data accumulated over a period from September 3, 2003 through January 16, 2004. The patch of sky in which the galaxies reside was chosen because it had a low density of bright stars in the near-field. (NASA, ESA)

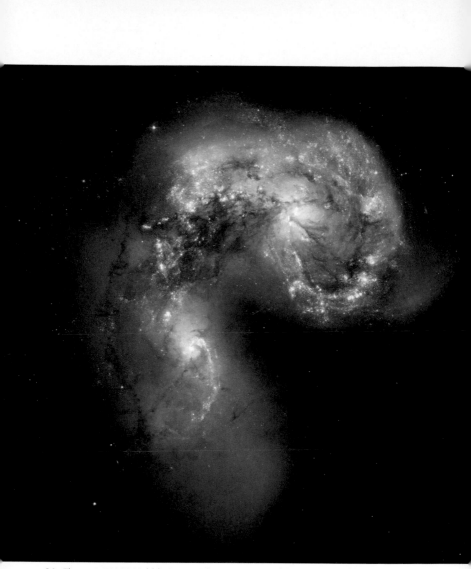

21. This new NASA Hubble Space Telescope image of the Antennae galaxies is the sharpest yet of this merging pair of galaxies. During the course of the collision, billions of stars will be formed. The brightest and most compact of these star birth regions are called super star clusters. (NASA, ESA, Hubble Heritage Team (STScI/AURA)-ESA/Hubble Collaboration)

22. The gravitational field surrounding this massive cluster of galaxies, Abell 68, acts as a natural lens in space to brighten and magnify the light coming from very distant background galaxies. In this photo, the image of a spiral galaxy at upper left has been stretched and mirrored into a shape similar to that of a simulated alien from the classic 1970s computer game *Space Invaders*. A second, less distorted image of the same galaxy appears to the left of the large, bright elliptical galaxy. (NASA/ESA; N. Rose)

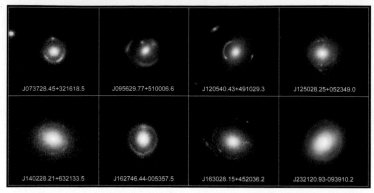

J073728.45+321618.5 J095629.77+510006.6 J120540.43+491029.3 J125028.25+052349.0

J140228.21+632133.5 J162746.44-005357.5 J163028.15+452036.2 J232120.93-093910.2

23. The blobs are giant elliptical galaxies roughly 2 to 4 billion light years away. The bull's-eye patterns are created as the light from galaxies twice as far away is distorted into circular shapes by the gravity of the giant elliptical galaxies. Gravitational lensing occurs when the gravitational field from a massive object warps space and deflects light from a distant object behind it. (NASA, ESA, SLACS Survey team: A. Bolton (Harvard/Smithsonian), S. Burles (MIT), L. Koopmans (Kapteyn), T. Treu (UCSB), L. Moustakas (JPL/Caltech))

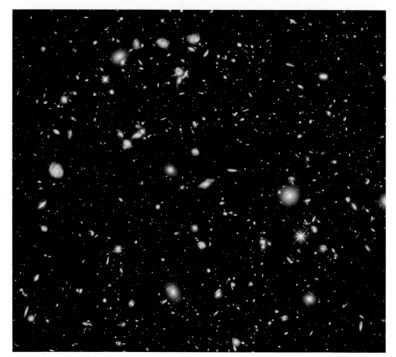

24. Called the eXtreme Deep Field, or XDF, this photo was assembled by combining ten years of NASA Hubble Space Telescope photographs taken of a patch of sky at the center of the original Hubble Ultra Deep Field. The XDF is a small fraction of the angular diameter of the full Moon. (NASA, ESA, H. Teplitz and M. Rafelski (IPAC/Caltech), A. Koekemoer (STScI), R. Windhorst (Arizona State University), Z. Levay (STScI))

by the rebounding material near the center of the star. But a supernova is more than just the rapid expulsion of the majority of the star.

During the initial collapse, as the very core of the star is pushed past white dwarf densities, the electrons and protons which make up an atom are compressed to such an extent that they form neutrons. Neutrons can sit much closer to each other than protons and their respective electrons, and so there's a new, stable end-point created here, as long as the exploding star is less than about 40 times as massive as our own Sun. This is the genesis of a neutron star. The name is simply descriptive – neutron stars are entirely made of neutrons. They are catastrophically dense cosmic entities. The process of shoving an electron and a proton together also produces an extremely tiny, extremely speedy particle called a neutrino.

A neutrino is a fundamental particle of our universe. It is neutral, as the name implies, and has extremely little mass. Neutrinos very rarely interact with matter (other than via gravity), they travel the universe at very close to the speed of light, and they're formed in a number of places. Our own star forms some neutrinos in its core. Because the neutrino doesn't interact very much with the stuff of stars and planets, once a neutrino is formed, it's very likely to escape the star almost instantaneously – unlike the poor photon, which will spend hundreds of thousands of years wandering the radiation zone.

Neutrinos may be generated by our Sun, but they're produced in *much* larger quantities during the collapse of a supernova and creation of a neutron star. Their instantaneous departure from the star means we on Earth can use them as an early warning system for nearby supernovae. The light may be still trapped within the star during this early collapse, as the rest of the star hasn't been blown outwards yet, but the neutrinos have already escaped. The time delay before the light can escape is not necessarily very long – for Supernova 1987A (color plate 15), it was only a few hours.

On top of the neutrinos flying in all directions, there's another tiny particle zipping around in the aftermath of the explosion – the electron. The material which isn't crushed downward into the neutron star is still made of standard old atoms, made of their respective protons, neutrons, and electrons. Typically the electrons are either bound to a specific atom, or at least floating around in the same material. But with the amount of energy that was flung through the rest of the star, blowing it apart, the electrons can absorb enough of that energy to go their own way. Traveling at close to the speed of light, they can produce X-rays if they are tangled around a magnetic field. The glow of these electrons, moving at fractions of the speed of light, can be detected by X-ray telescopes around our Earth.

In general, it's hard to predict exactly when a given star will cross its internal threshold to begin its collapse and supernova outburst. This unpredictability is one reason we like the neutrino alarm mechanism – if we detect a burst of neutrinos, we have a little bit of a heads-up that something is coming. Our best method of spotting lots of supernovae is still to survey the sky repeatedly and look for changes – something that upcoming telescope facilities like the Large Sideral Synoptic Telescope (LSST) will excel at. Occasionally, however, we find a nearby sibling of our Sun that looks like it might go off soon (on an astronomical timescale, of course). Such is the case with Betelgeuse.

Betelgeuse is due for an explosion

Betelgeuse is a red supergiant star, and we've mentioned already that it's the left-hand shoulder of the Orion constellation, a visibly orange star in the night assembly. It is, in fact, one of the top ten brightest stars in the night sky. (The Sun is somewhat obviously always the brightest object in the sky.) Betelgeuse is also one of the few stars that's close enough for us to resolve in more detail than a point source of light.

At a distance of 600 light years from the Earth, if Betelgeuse were to go supernova right *now*, we'd clearly have a 600-year delay before we could spot any changes from here. So whenever it explodes, it's going to be a long wait before we see it, but in the meantime I can do some math to work out how bright Betelgeuse's explosion will be, albeit making assumptions about the exact way it explodes. The precise flavor of this supernova is still up for a bit of debate, depending on exactly how fast it's spinning and how quickly it gets rid of its outer layers over the next 100,000 years. But all the supernovae options for this star reach about the same brightness, so for a quick calculation, we don't really need to worry about the exact type of explosion this star will undergo.

There are two ways of measuring brightness in the astronomy world. The first is absolute magnitude, which is the brightness of the star as it would be measured from a fixed distance. (It's arbitrary, but the fixed distance chosen is 10 parsecs, or about 33 light years.) This is trying to get to a measure of intrinsic brightness – as though we could line up all the objects in the sky at equal distance from us, and compare them to each other. We can't actually *measure* the brightness of a star this way, but we can apply some corrections based on the distance to the star in order to get to it. The absolute magnitude of a Type II supernova is around −17. Because astronomers have the worst conventions in the world (for largely 'historical reasons'), negative numbers mean brighter objects. The Sun has an absolute magnitude of 4.83, which, once we translate out of 'magnitudes', means that the Sun is 500 million times fainter than the −17 supernova, when measured at the same distance. This huge difference in relative brightness shows why a supernova can outshine an entire galaxy.

The other method of measuring brightness is a bit more straightforward. It's the apparent brightness − i.e., how bright it appears to us as viewed from the Earth. In this frame of reference, more distant objects will always appear fainter, regardless of how *intrinsically* bright they are. Because Betelgeuse is still fairly distant from us, the apparent brightness will be significantly less than the absolute magnitude. Based on the distance to Betelgeuse, we can work out that the apparent magnitude of the peak of the explosion will be −10. The Sun, the brightest thing in our sky, checks in at an apparent magnitude of −26.74. Once again translated out of magnitudes, this means that the Sun as seen from the Earth is a whopping 5 million times brighter (more or less) than Betelgeuse's explosion, so this supernova certainly won't be anywhere near as bright as our Sun in the daytime. That's not to say you wouldn't be able to see it − it would definitely be bright enough to see during the day, as long as you were looking in the right direction. (After all, as we noted before, you can still see Venus in the daytime, if you know where to look!)

Nighttime will be a different story. The brightness of Betelgeuse's supernova will be about the same as the quarter Moon. It will also be about sixteen times brighter than the brightest supernova known to have been seen from Earth, which occurred in the year 1006 and was recorded by a number of early civilizations. It was said that the supernova in 1006 was bright enough to cast a shadow at night. Betelgeuse, being significantly brighter, will likely also cast shadows – which, if you think about the brightness of a quarter Moon, would make sense.

Betelgeuse isn't expected to explode for another 100,000 years or so, but we do expect a few supernovae in our galaxy every couple of hundred years, so there are a number of stars that are nearing the ends of their lifetimes within the Milky Way.

When your stellar sibling dies before you

We mentioned earlier that not all stars are as isolated as our Sun. Stars can gather together in tight globular cluster families, but there's another common configuration – binary stars. These are stars that formed close enough to each other that they are gravitationally tied together, and will travel the length of the galaxy like this. These stars are twins, of a sort, though most of them will be fraternal, not identical twins. About a third of the stars in the Milky Way are in some kind of binary, though this accounting includes all the stars which are only very gently tied to each other. If you want to include only close binaries, where the two stars orbit each other in years instead of hundreds of years, the number drops down considerably.

Since the majority of these stars aren't identical twins, they won't have exactly the same mass. If their masses are different, then one star (the more massive of the two) will turn into a red giant before the other. It will expand outward into the space between the stars, and the second star, which is still burning hydrogen in its core, may begin to tug on the outer layers of its red giant twin, slowly pulling them in towards itself.

The red giant, feeling this pull, will no longer look like a giant fuzzy sphere, but the side of the star facing its twin will be pulled away sharply into more of a fuzzy teardrop shape.

What the companion star does with that surface material depends on what kind of star it is. If the companion to the red giant is still burning hydrogen in its core, it should be relatively stable on its own. But with an influx of gas from its red giant twin, the stable twin will find itself unexpectedly growing in mass. If you grow a star that's already able to burn hydrogen, you've added to its gravitational inward force, meaning that the temperature at the core of the star can be increased, and the speed with which it can burn through its hydrogen also increases.

What of the red giant? Well, it was going to lose those outer layers anyway – donating them to its twin is the fastest recycling job it could have done. And it won't donate itself into oblivion. Its twin star can only pull the layers which are furthest from the red giant's core; the hot dense core of the giant is staying right where it started. This setup can last for a long time; the red giant, slowly expanding, gradually feeds more material directly into the gravitational weight of its twin star. That twin star, by growing in mass, only shortens its own lifetime, burning hotter and faster than it would otherwise, had it not had a twin star to keep it company.

After the explosion

When we dust off our post-red giant bursts, we're largely left with low-mass stars that have produced planetary nebulae and only remain as white dwarfs, and high-mass stars that have turned into supernovae, leaving only neutron stars behind. Both white dwarfs and neutron stars have been found in pairs with other, evolutionarily younger stars, a role-reversal of our fraternal twins from above.

In these situations, instead of a red giant and a hydrogen-burning star, we have a hydrogen- or helium-burning star and a white dwarf or neutron star. Let's start with a red giant, our Sun's helium-burning stellar sibling, paired off with a white dwarf. Much like the scenario we outlined earlier, the white dwarf, if it is close enough, will pull material away from the red giant, gradually adding to its own mass. Unlike the hydrogen-burning star, though, a white dwarf is doing no burning in its core. All that happens is that its mass grows.

White dwarf stars are not terribly tolerant of gaining weight. Eventually, this white dwarf will trigger an explosion of its own. The less destructive option is for a thermonuclear detonation to occur on the surface of the white dwarf. This surface detonation is called a 'nova' – like the supernova, it is a dramatic, bright flare. Unlike the supernova, it is not so bright that it will outshine the galaxy it lives within, and it doesn't destroy the white dwarf. So the white dwarf, with a steady source of fuel, can build up enough mass to trigger a detonation, blasting the material away from its surface, before settling down to build up more mass, triggering yet another surface explosion. This kind of behavior makes these stellar twins fairly noticeable, because the brightness of the star will flare to many times its original brightness.

The other option is for the white dwarf to undergo a supernova. The white dwarf itself is only stable up to a mass of 1.4 times the mass of the Sun. Beyond that, and you gravitationally compress the star beyond what it can stably support. The gravitational crushing forces for a white dwarf can very briefly press the star to a temperature where it can burn carbon. And so it does, catastrophically, and all at once. The energy unleashed by the whole star undergoing a giant simultaneous burn flings the star apart, leaving nothing behind. This style of supernova is unique to white dwarfs – the higher-mass stars have already burned through all the carbon at their cores.

If, instead of a white dwarf, the twinned star is a neutron star, we wind up with what's called an X-ray binary, for the somewhat boring reason that it produces a lot of X-rays. As the gas from the red giant (or a hydrogen-burning star) is pulled away, it gets stretched into a very thin disk, surrounding the neutron star. The disk forms because it's very hard for gas to lose a lot of momentum all at once and plunge straight down onto the neutron star. Because the disk is there, the gas is heated up to an incredibly high temperature before it makes it all the way to the neutron star – or to a black hole (see below). This heat causes the X-ray glow we can observe, and keeps the disk itself almost invisible in optical light.

No matter what kind of supernova or nova explosion a star undergoes, these catastrophic, extremely energetic events have no impact on the galaxy as a whole. Even if a star entirely self-destructs, it only makes up one trillionth of the mass of its galaxy. That explosion, even though it can outshine the galaxy, is simply not big enough to have a noticeable impact – one star can only put out so much energy. The outshining can be dramatic in images of the galaxy, however – a bright star can suddenly appear, and then gradually fade from view.

There is a critical role for supernovae in a galaxy, though. Without the supernova, we'd only have elements up to iron to work with. During the ricochet of the outer core of the star, the heavier elements are formed, giving us gold, silver, and platinum, for instance. These elements are scattered along with the remains of the star into space. When the next generation of stars forms, they will incorporate those metals into themselves, carrying forward the remains of generations past. The Earth's supply of these precious metals, along with all the other elements on the periodic table beyond iron, come from previous generations of our Sun's siblings catastrophically exploding and spreading these elements throughout the galaxy.

There's one outcome of a supernova we've skipped over so far. If the star is massive enough, the collapse of its core won't stop at a neutron star. Gravity will continue to pull each neutron closer to its neighbors. Unfortunately, there's no new particle that neutrons can collapse into when compressed too far, and what happens next is mathematically difficult to describe. At this point, the runaway collapse *should* continue. With no resistant force, it can become infinitely small, with an infinitely deep gravitational well. This object, infinitely small, infinitely dense, is termed a singularity. Infinities and physics currently do not play well together, so our descriptions of how space and time behave immediately surrounding the singularity begin to fail. (The singularity itself is currently most accurately described by a large number of question marks.) This collapse has produced a new astrophysical object – it has become a black hole.

Black holes

Black holes are some of the most interesting remains of our Sun's sibling stars, and they were first suggested by people who were tinkering with our understanding of how gravity works. Black holes, at the time, were completely theoretical additions to our cosmic family. In 1784, John Michell realized that if you had a sufficiently compact, massive object, the escape velocity – the speed needed to escape from the surface and make it out into orbit and beyond – would be faster than the speed of light. Since nothing moves faster than the speed of light, these objects become a trap for anything that comes close enough to get dragged in.

The term 'black hole' has expanded a bit to mean more than just the very 'object' of the black hole, that infinitely small singularity. This is reasonable enough, because there is a region of space surrounding the black hole where the gravitational pull of the singularity is so strong that light itself would be captured, which is marked by an imaginary line called the event horizon. This event horizon is usually quite distant from the actual singularity, but the term 'black hole' is often slung around for this entire region, because it is inside this space that the black hole's reign is complete.

It takes a lot of evidence to get us from a theoretical member to a confirmed member of our stellar family, but fortunately over the years we've built up just such a pile of evidence for their existence.

Some of the best evidence we have for adding black holes to our cosmic family tree comes from the center of our own galaxy. With the telescopes we have now, we can watch individual, luminous stars zip around *some* extremely heavy, physically small, and completely invisible object. From the orbits of the stars, we can figure out how much matter the dark object must contain. We also know how big it could be, since some of these stars pass very close to the mystery object, and aren't being torn to shreds, so it must be smaller than that. With the mass and the size, we can have a guess at its density.

The object at the center of the Milky Way is so dense that no other known or suggested object from anywhere in our family tree, aside from a black hole, can explain our observations. The object is about 4 million times the mass of the Sun in a space that's less than the size of our solar system. The only way you can pack material that tightly is if you crush it to the point where the object has no other physical option than to be so dense that light cannot escape it: it *must* be a black hole.

We know that other galaxies have black holes too, so ours isn't weird or special in some way by having this massive black hole at its center. In fact, as far as we've been able to tell, *every* galaxy has a black hole at its center. If you're looking at a galaxy that's relatively nearby, you can tell how big its black hole is by how the galaxy is rotating. If you want to be able to model the way it rotates, you need to have a good understanding of where the matter in the galaxy is. Generally we can find out where most of the matter is, because a good chunk of it is either in gas (which we can see) or stars (which are even easier to see). However, if we don't also include a whole pile of extra matter right at the center (like a black hole), the models won't reproduce what we observe. If we stick a black hole in there, we get a much better match.

Similarly, we know that there are smaller black holes hanging around in our galaxy – supermassive black holes aren't the only ones in the universe. Much like we see stars zipping around an invisible, apparently massive object in the core of our galaxy, we've seen a huge number of stars which appear to be orbiting an invisible companion closer to their own mass – the twin stars arrangement again, but instead of a white dwarf or a neutron star, their companion is entirely invisible. We can repeat the density calculations we did before, and we find that again, we're dealing with an object that is sufficiently tiny and sufficiently massive that only a black hole fits the requirements. Our family has a new member – the black hole.

Gravitational wells

Black holes are a particularly imagination-inspiring member of our cosmic family, but their role as a light-trap, along with the difficulty of observing them directly, has created some myths about them. One of these is the idea that a black hole is some kind of universal-scale vacuum cleaner, constantly sucking material inwards towards itself.

This point of confusion seems to come from a few sources. One is that we often describe the force of gravity as one that 'pulls' or 'attracts' two objects together. If you combine that image with the idea that a black hole can be considered like a cosmic garbage bin (throw anything in, nothing comes out), then it seems very reasonable to think that not only are they inescapable light traps, but that they pull things towards them as well. Unfortunately, black holes do no pulling whatsoever, as we'll see.

When we talk about the gravitational influence of an object, it's often described in terms of bowling balls on rubber sheets, or like the image below. The more massive the object, the deeper the indentation in space becomes, and the steeper the indentation finds itself.

These are good and reasonable explanations, though they are simplified since the indentations happen in all directions, so these 'indentations' are really three-dimensional distortions, or condensations, of space itself. These simplifications do illustrate an important point – fundamentally, gravity operates like a distortion to space. Often we speak of this distortion in terms of a 'well', which

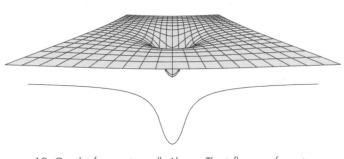

12. Graph of a gravity well. Above: The influence of gravity on the surrounding space. Below: Gravitational potential well.* (Wikimedia user BenRG, public domain)

goes back to this idea of a two-dimensional sheet bent into a third dimension. It's easy to think of things rolling downhill into a divot, formed by the presence of a large amount of mass.

Mathematically, the influence of gravity is written out as *directly* proportionate to the mass of the object, and *inversely* proportional to the square of the distance between you and that object. The direct dependence means that as you increase the mass, you increase the influence of gravity by the same amount. The inverse dependence means that as you double the distance, the influence of gravity gets

....

* The gravitational potential is a way of quantifying how much energy you would have to expend in order to lift something away from a massive object. The deeper you are in the well, the more energy is needed, and the steeper the well, the faster you'd need to go to escape on a rocket.

four times as weak. This strong dependence on distance is what creates the particular smooth curve that we draw to illustrate gravity's influence from the center of the object, as in the lower part of the diagram above. If you're very close to the massive object, the gravitational well is deep, and you feel a strong gravitational force. If you're further away, gravity can't distort space very much, and so you feel a much weaker gravitational force.

All objects in the universe create their own distortions to space, including our human bodies. However, the force of gravity is relatively weak, and we humans are not very massive, so we don't expect to see objects around us rolling into our gravitational well. Once you scale up to moon or planet amounts of mass, then the force of gravity means they can have a little more influence on their surroundings. Each of the planets thus has its own distortion to space, but all of these distortions pale in comparison to that caused by the presence of the Sun.

Critically, all of the distortions caused by massive objects didn't spontaneously appear when the object finished forming. Each object simply had a gentler gravitational well before it finished collapsing into a denser object. A small gas cloud could have the same mass as a small planet, but the planet's gravitational well will be steeper in the center than the gas cloud's. If you kept compressing the planet's mass until it was dense enough to be considered a black hole, its gravitational well would be steeper yet in the center. The steeper the gravitational well, the faster you need to go to escape it.

The event horizon of a black hole, in this context, describes the boundary within which the black hole's gravitational well is so steep that not even light could escape it. If you take one of the rubber sheet diagrams, you could place down a circle to describe the event horizon's location, though in the universe these are actually spheres in three dimensions. The event horizon itself doesn't describe a physical boundary to the influences of gravity. The gravitational distortion itself exists continuously outside and inside of the event horizon – outside the event horizon it's just slightly less steep. You could place another circle outside the event horizon that describes the location at which you need to go half the speed of light to escape – again, there's no change or boundary in the physical distortion at this circle, it's just a descriptive line that might prove useful to understand or describe the object's influence on space.

If you're far enough away from the black hole, then the influence of gravity from that black hole is no different from having a star of the same mass – or a planet, if you somehow got yourself a particularly tiny black hole. If you replaced our Sun with a black hole of the same mass, the planets would continue to orbit the way they always have; their orbits are determined entirely by the mass of the object at the center of the solar system, and not by that object's density. If our grandparent star couldn't pull an object into its atmosphere, the black hole would struggle to do so as well.

Black hole horizons

If we use the event horizon of the black hole to describe the 'surface' of the black hole, even though there's no surface there, we can start to describe the geometry of this sibling of our Sun. We mentioned earlier that the event horizon is usually a circle when we simplify the gravitational pull down to two dimensions, so in three dimensions, the event horizon should be a sphere. In fact, black holes are probably some of the most perfectly round objects in the universe.

In principle there's no up, down, or sideways to a black hole, just as there's no up, down, or sideways to a spherical toy. If the black hole is totally isolated, this is the case. Any direction you look at it, it will still appear as a circle of utter blackness, where no light is produced, reflected, or passing through. The only change you'd see by looking at it from different angles would be the backlighting. The backlighting of the black hole isn't *just* the background light; black holes bend light that passes near them, and so all the light from the stars behind the black hole can end up compressed into a ring of light, surrounding the black hole itself. The more objects behind the black hole, the brighter that ring of light will appear. If you found a direction with very few stars behind the black hole, this ring of light might seem to vanish.

The exception to this is if our black hole is one of a set of twins, alongside a companion star. Much like the white dwarf star and the hydrogen-burning star, the black hole will start to siphon gas off of the surface of the other star if the star's atmosphere expands outwards and drifts too near. While it's one thing to pull the gas away from the other star, it's quite another to pull that material inwards past the event horizon of the black hole. The gas from the other star has to lose a lot of energy in order to fall into the black hole, and so the black hole winds up wrapped in a many-layered blanket of the other star's gas. This is called an accretion disk, as the many windings of the other star's gas form into a disk shape around the black hole (see color plate 16). Accretion is simply the black hole's attempt to gain that material for itself, and grow its own mass in turn. Black holes are very, very bad at getting material that's handed to them like this all the way down past the event horizon, so the accretion disk is usually relatively stable, and the black hole doesn't actually grow in mass very much.

However, thanks to the presence of this accretion disk, there's suddenly a direction, or an orientation to the black hole. You could look at the black hole 'from the side', which would be looking at the edge of the accretion disk. Or you could look down on the disk 'from the top', where you'd see the black hole, plus the glowing disk of gas surrounding it.

There are a few other situations in which you might see some weird things around a black hole, and where the spherical symmetry is broken. One of them is if two black holes are orbiting each other. You could get this scenario from a twin star setup, where both of the stars were massive enough to explode in a supernova and create a black hole in their aftermath, but without throwing their twin out of its orbit. Light bends around a single black hole in a pretty dramatic fashion (see color plate 19); if you add a second black hole in close proximity to the first, the dance that light can get caught in is an extremely elaborate one. This convoluted dance means that only one of the black holes actually looks spherical; the other one's shape is distorted, as the ring of backlighting is also distorted, and the black hole nearest to us ends up with a small curved arc of darkness hovering near it. This has technically been called an 'eyebrow'. They're also referred to as 'black hole shadows' – which is just as great, name-wise.

Black hole collisions

These twinned black holes are quite stable, and capable of lasting billions of years without much evolution. However, if their orbits are less circular, or the black holes are not the same mass, or enough time has passed, you can catch them colliding.

Black holes go through a long spiraling wind-down before they actually collide, and the distortions they make in space as they circle each other create gravitational waves in space-time, very like what happens if you swirl your finger in a pond. The formation of these waves carries energy away from the two black holes, helping them lose enough energy to merge together. Like sound waves, gravitational waves have a frequency and amplitude. These are at extremely long wavelengths, so it's nothing the human ear could hear normally, but if you scale the frequencies up by a couple of thousand, you can get an impression of what it would sound like. This scaled sound is often described as a chirp, but if you listen to one, it sounds much more like a *vwooop*.

Making these noises isn't just a curiosity – there's a real

scientific reason to create them. As the Laser Interferometer Gravitational Wave Observatory (or LIGO for short) goes hunting for the signature of gravitational waves in the universe, the collisions between black holes are one of the first things we should be able to detect – they're one of the 'louder' sources of gravitational waves. As with any detections, it helps to know what you're looking for, and these predictions of what black holes of a specific size should do to the space surrounding them as they merge are key to developing automated methods of spotting these gravitational waves, as they wash over our planet.

On February 11, 2016 LIGO announced the very first detection of the gravitational wave signature of two black holes colliding. Below is their illustration of the gravitational wave arriving at both of their facilities. With a large set of predicted *vwooops*, the LIGO team found that their particular chirp was closest to a model of two black holes, one 36 times more massive than our Sun, and one 29 times more massive, reaching the end of a long spiraling orbit, and finally colliding.

In this instant of merging, the event horizons of the two black holes cross, and we can consider them as a single object. After all, there's now one contiguous region where light can no longer escape. This is another moment where black holes are not perfect spheres.

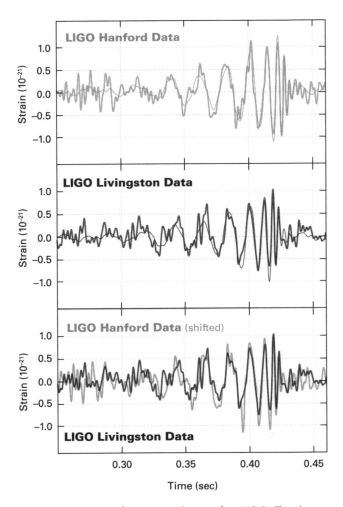

13. A measurement of gravitational waves from LIGO. This shows the gravitational wave signals received by the LIGO instruments at Hanford, Washington (top), Livingston, Louisiana (middle) and combined (bottom). The faint line on the top two shows the pattern predicted by the model of inspiral black holes. (LIGO)

The newly formed larger black hole will quickly become reasonably round-*ish*, but it has to go through a fairly long process called 'ringdown' before it can completely settle back down into its earlier, perfect sphere shape.

If ringdown sounds like a melodic term to you, you're spot on – it's meant to evoke the gradual dimming of sound that comes from a bell after being struck. The bell continues to vibrate long after the initial strike, carrying a tone at a quiet level much longer than you might think. So it is with our Sun's black hole cousins. After the initial strike of its twin black hole, the conglomerate black hole continues to vibrate for a long time, gradually losing energy to space until it finds itself back the way it started – an isolated sphere marking a region of gravity's absolute dominance.

Black holes and the passage of time

Black holes are also rolled out whenever scientists wish to discuss something odd that happens when space or time behave in an unusual fashion. It's the bending of light in strong distortions of space that gives black holes their odd backlighting, but as space and time are bound together as space-time, whenever space is doing something odd, time must be too.

Within Einstein's general relativity, there is a concept

called time dilation, which implies that our perception of how quickly time is passing depends on a combination of how fast we're moving and how distorted is the space in which we're sitting. Space is distorted near very massive objects (like black holes) and so if we're either moving quickly (relative to the speed of light) or near a black hole, our perception of time might differ from the perception of time that someone else, who's not moving quickly or near a black hole, would measure.

Movies like *Interstellar* brought the idea of time dilation a little further into the public awareness, and they actually did a pretty good job of explaining what was happening, especially considering how complicated anything to do with relativity can get. Time dilation is a property of our universe, though, and it's apparent on much smaller scales than the black hole of the movies – we can easily measure its role around our own planet, the Earth. If you've ever used the 'my location' option on your favorite map app on your phone, you've made use of it yourself.

The principle of time dilation is this: the deeper you find yourself in a strong gravitational field, the slower your clock will run, *relative to* someone who is not in as strong a gravitational field. This can be translated into meaning the closer you are to something large, the slower your clocks will run. However, it also means that if your two clocks are the same distance from two objects, one which has a much stronger gravitational pull than the other (say, a planet for one clock, and a black hole for the other), the clock around the more massive object (the black hole) will run slower, even though both clocks are the same distance away from their respective objects.

A GPS satellite, which orbits about 12,500 miles above the surface of our planet, is clearly further away from the Earth than those of us who live on its surface. A surface-of-the-Earth clock will therefore appear to progress more slowly, relative to the clock on the GPS. Unfortunately, we really need these two clocks to operate in sync with each other, otherwise the location information you get back from the satellite starts to be increasingly inaccurate, defeating the point of having the GPS in the first place. This effect can be mathematically calculated, so we can correct for this slight difference in clock speed by setting the GPS clock to run a little slower than normal. Just a little slower – this effect is only nanoseconds in size near the Earth.

But for a GPS, another form of time dilation is also in effect, because the fact that the satellite is in motion changes the onboard clock as well. This form of time dilation is such that the faster you're moving, the *slower* your clock appears to move, relative to someone who is not moving. This effect gets more and more dramatic the closer to the speed of light you travel, so the offset between your clocks would get more severe.

A GPS satellite, which has to travel in a circle 12,500 miles above the Earth's surface every 12 hours, is therefore going around 8,670 miles per hour. The speed of light is about 670,000,000 miles per hour, so our GPS is only going about 13 millionths of the speed of light. This is not a significant fraction of light speed, but it's enough that we can calculate and measure the impact that it has on the onboard clock on the GPS.

As it happens, the gravitational time dilation has a more significant impact on the clock than the speed of the GPS relative to us on the ground. While the fast running of the clock due to gravity is partly canceled out by the slow-down imposed by its 8,670 mph speed, the clock is still running a bit fast because it's so far away from the most concentrated part of the Earth's gravitational field. It's this slightly diluted rapid ticking of the clock that's calibrated into the clocks that we send up onto our satellites for the GPS network.

If you're in a lower orbit, as the astronauts on board the International Space Station are, the gravitational effect isn't a big one, so the only clock-altering effect they deal with is due to their orbital speed around our planet. If you spent an entire year on the ISS like astronaut Scott Kelly, you would have aged a grand total of 0.007 seconds less than your family on the ground. This is an extremely small number, not because the ISS is going particularly slowly (it orbits at around 7.7 km/s), but because the important factor is how close to the speed of light you're going, and for the ISS, the speed of light is *so* much faster – about 39,000 times faster, in fact.

If you were to put yourself in orbit around a black hole, where the gravitational distortions are much more extreme, all of these effects would also be much more extreme. An orbiting craft around a black hole, if the black hole were large enough, could absolutely feel time running considerably slower than a spacecraft which hadn't approached anywhere near it.

Anytime you pull your phone out to get directions to a restaurant, or to check how far you have left to walk, and your phone accurately figures out where you are, the GPS connection that underlies that software is relying on our understanding of relativity, and the time dilation the satellite is undergoing, to do its job properly. The deeper and steeper the gravitational well, the more important these accountings of time dilation are, so if we were exploring a black hole or a neutron star, these corrections would be critical!

THOUGHT EXPERIMENT:
What happens to time dilation
at the speed of light?

There's another fun side to time dilation. If time dilation, which is a stronger effect the faster you move, is present for all moving objects, how should it affect light itself? After all, light is the fastest object in our universe.

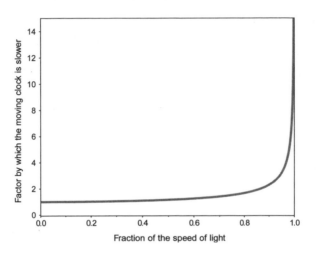

14. This graph shows the relative slowing of a clock in motion, as a function of how fast the clock is moving. The closer to the speed of light the clock is moving, the slower it appears to tick relative to a stationary clock. (Image courtesy of the author)

To answer this, we have to understand a bit more about *how* the clocks change as you increase in speed. If you put one person on a very fast spaceship and send them to go explore the cosmos, and keep another one at home to wait for their reports, the faster the ship goes, the more discrepant their clocks will be. The factor by which the clocks are out is called the Lorentz factor, named after Hendrik Lorentz who mathematically described how this works.

This factor is also what's plotted on the vertical axis of the figure above. Along the horizontal axis is speed, so the faster you get, the more out of line your two clocks are. The faster and faster you go, this difference increases almost exponentially. It's not actually an exponential, but it's close in shape. So the closer you get to the speed of light, the more dramatic the discrepancy between the traveler and the Earthbound person becomes. The closer to the speed of light you travel, the longer and longer the Earthbound observer will wait for one of your seconds to pass.

At the speed of light exactly, time dilation gets a little weirder. The exact relation magnitude of the difference between the fast traveler and someone back home on Earth is determined by the ratio of one over the square root of the following:

$$(1 - [\text{velocity}^2 / \text{speed of light}^2])$$

If the velocity is equal to the speed of light (as it would be, for our imaginary photon hitchhiker), then the second part of the equation, in square brackets, is equal to 1. Now subtract 1 from 1, and you get zero. The square root of zero is still zero, and then we run into a problem, mathematically, because we have to divide 1 by zero. Any number divided by zero is infinity. This is why the Lorentz factor suddenly shoots off the top of the figure above when you reach the speed of light.

According to our current understanding of this physics, a person on Earth would have to wait an *infinite* amount of time to observe any amount of time passing on the photon. As far as the Earthbound observer is concerned, time has stopped for the photon at the speed of light.

This doesn't mean that photons are zipping around, instantaneously bashing between objects. Light still has a finite speed – and that's a real measurement of the physics of the universe. Our determinations of light years are still good, impartial, measuring sticks for the universe.

✳

With supernovae, planetary nebulae, and black holes representing the end of the stellar life cycle, we've seen stars in all their luminosity and in their vigorous finales. It's time to look outwards once more, to the galaxy which hosts all the stars and the remnants of stars in our neighborhood – our own Milky Way. Just as our home star, the Sun, was the first entry into a vast family of stars, so the Milky Way is our portal into the vast family of galaxies which stretch throughout the universe.

6

GALAXIES

The galactic family

If you look up at night, far away from the light produced by cities, you can see that our stellar family isn't scattered randomly across the whole sky. There is a clustering of stars, so dense that it might at first appear to be a wispy, high-flung cloud, which stretches across the entire sky in a band.

This band is the next step up in our cosmic family tree – it's our home galaxy. Taking the family tree approach, this galaxy, the Milky Way,★ is our great-grandparent. Within the Milky Way, all of our grandparent stars are held together, bound by their mutual gravity and the gravity of diffuse gas and dust – the marker of other stellar grandparents which have already come and gone, and the creator of new, future generations of stars. In the early universe,

★ The word galaxy comes to us from the Greek 'galaxias', which translates to 'milky', after the appearance of our galaxy as seen from dark skies.

it was the coming together of gas, dust, and dark matter*
which drew in ever more material, fostering the growth of
a gigantic stellar family within each galaxy's gravitational
bounds. Much as the gas which forms stars collapses downwards into a disk of material, spinning more rapidly than
it had been before that collapse, galaxies are also rotating
disks of material – on a colossal scale.

A galaxy is a collection of gas, stars, dust, dark matter,
and at least one large black hole, which sits at its center,
and our Milky Way is no exception. To try and place our
galaxy among its sibling galaxies, we examine their size,
shape, and color.

Unlike stars, which we divided up primarily based on
their color, a galaxy's first division is based on shape. The
vast majority of galaxies come in one of two shapes. They
are either shaped like a slightly deflated ball – roundish
along one direction, a little longer in the other direction,
and fuzzy all over – or like a pancake, an extremely flat,
round object, with spiraling concentrations of stars winding
outwards from their centers. The fuzzy, deflated galaxies are
called ellipticals, a term meant to describe their oval shape.

---

* Dark matter is a still mysterious component of our universe. It seems to
make up about a quarter of all the material in the universe, but it interacts
with the matter of stars and planets only through gravity. Many searches
for the particle responsible for dark matter are ongoing!

The flat galaxies are called spirals, in turn meant to describe their swirling patterns (see color plate 17).

Edwin Hubble was one of the first people to try to classify galaxies, and he developed a system that we've since dubbed the 'tuning fork diagram' (see below), which puts the smooth, featureless ellipticals on one side (usually the left). The fuzzy elliptical galaxies are mostly families of very old stars, and are thought to have formed early on in the universe. We think that they might also have been flat spiral galaxies at an earlier point, but that over time, the stars

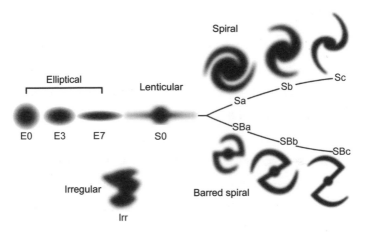

1.5. Galaxies are fundamental building blocks of the universe. Some are simple, while others are very complex in structure. As one of the first steps towards a coherent theory of galaxy evolution, the American astronomer Edwin Hubble developed a classification scheme of galaxies in 1926. (NASA, ESA)

within that galaxy got pushed into more and more random orbits, gradually making them rounder.

Spiral galaxies go on the other (usually right-hand) side of the diagram. Their structure is a bit more complex than the fuzzy, amorphous ellipticals. The flat portion of the galaxy is called the disk, with a central, fuzzier (and less flat) portion in the very middle, dubbed the bulge.

The two tines of the Hubble 'fork' diagram mark the next tier of our sorting of galaxies. There exists a set of spirals which have a rather odd structure in their centers – what appears to be a straight line crossing the very center of the galaxy. These structures are called bars. They can be such a dramatic feature of a galaxy's shape that we separate those galaxies which have a bar from those without. Since these are all galaxies with the same general shape (flattened, with winding spiral 'arms' radiating outwards from their center), they retain the name 'spiral' galaxy, but to distinguish them they're called 'barred spirals'.

It's common for barred galaxies to be drawn with only two arms, but this isn't always the case. We have determined that our own galaxy has a relatively strong central bar, and well more than two spiral arms. In fact, if you look at our best maps of the Milky Way, not only do we have more than two spiral arms, but we also have smaller branches (sometimes labeled 'spurs') between the arms, as the arms themselves fork and divide. It's within one of these little offshoots that our solar system – with Sun, Earth, and every human – resides, about two-thirds of the way out from the center of the galaxy.

There are also exceptions to this general set of elliptical/spiral/barred spiral galaxies, because nothing in the universe is completely straightforward. There's a class of galaxies which are (really, in all seriousness) technically called 'flocculent spirals'. These galaxies are still pretty flat – they're more like a spiral galaxy than an elliptical – but they're remarkable because they completely lack the spiraling arms. They have tufts and wispy bits, but nothing so dramatic as a spiral arm. (Flocculent, in case you are curious, means 'wool-like', which is a rather pleasing mental image for describing the disk of a galaxy.) Our nearest neighbor, the Andromeda galaxy, is not quite a flocculent spiral, but nonetheless doesn't appear to have strong spiral arms; it shows dark dust lanes instead.

All this said, that two-armed spiral pattern we so often associate with galaxies isn't incorrect – there are still plenty of galaxies out there with strong, definite spiral patterns of stars, with very little filling in the gaps between those arms. These galaxies have the delightful name of 'grand design spirals'.

Grand design spirals (which only make up about 10%

of galaxies) tend to have either two or four strong spiral arms, and they're often extremely photogenic and pleasingly symmetric objects in the sky (see color plate 18). However symmetric they may be, exactly how those spiral patterns are made and maintained has been remarkably hard to explain.

The problem with understanding spiral arms lies fundamentally in that the components of a galaxy rotate at different speeds; the inner part of the galaxy rotates faster than the outer parts. So the easiest explanation – that the arms are actually just areas of the galaxy that are physically more dense, and a fixed component – doesn't work. This is excellently demonstrated with the Sun's magnetic field, which gets increasingly tangled over time, as we've seen. The inner part of the galaxy will rotate many more times than the outer part will, so if the arms were physically planted, like the magnetic field is in the surface of the Sun, we would expect the bright arms to wind themselves into an incredibly tight spiral.

Any solution to this problem has to mean that the spiral arms can't be attached to anything, and they can't be a physical object. So far the best explanation is called Density Wave Theory, which is effectively what happens if you have a traffic jam going in circles. The spiral arm is a density wave, a compression of the material that already exists in the galaxy, but stars aren't fixed to a location 'inside' or 'outside' of the spiral arm. Like cars passing through a traffic jam, while the stars are within the arm, they're in a very tightly packed region of space (and therefore quite bright, as the light from a large number of stars adds together before streaming towards us), but they're not stuck there for ever, and eventually they will pass onwards, as the density wave moves past them.

There are a *lot* of galaxies

Just as the Sun is part of an astoundingly large stellar family, so our galaxy is also part of a universe-wide family of galaxies, separated by huge volumes of space. If we look around our home galaxy, it can be difficult to appreciate just how many other galaxies are out there. However, if we look beyond the Milky Way, we can capture a larger fraction of the universe at one time than we can if we're only looking near ourselves.

The Hubble Ultra Deep Field is one such image of the distant universe, and it gives us a flavor of just how galaxy-rich our universe really is (see color plate 20). I encourage you to find a high-resolution version online and look around. Every single bright pixel in this image is a galaxy (with the exceptions of the bright spots with spikes – those are stars in our galaxy that got in the way). This image alone contains 15,500 galaxies, and it's a tiny, tiny fraction of the night sky. It's about 1,200 times smaller than the area blocked out by the tip of your little finger held at arm's length.

The patch of sky imaged for the Ultra Deep Field was chosen because it was particularly dark, thus avoiding light contamination from our own galaxy. Certain parts of the sky are totally unusable for studies of distant galaxies; the gas and dust in our own galaxy is simply too bright in those directions. You'd also struggle to find anything directly behind a nearby galaxy. We'll want to be looking at the blackest, emptiest part of the night sky we can find.

The down side to choosing an extremely dark patch of sky is that the galaxies you're interested in are very, *very* faint. Even the best telescopes have to dedicate an extraordinary amount of time to capturing the light from these distant galaxies. For the Ultra Deep Field, the Hubble Space Telescope spent 1 million seconds staring at this patch of sky. If this had been observed in a single session, it would have lasted for a little over eleven and a half days.

If we could spend the same amount of time on every patch of sky, we'd expect to get a similar-looking image every time. This means we could take the Ultra Deep Field, with its 15,500 galaxies, and imagine tiling it out across the sky. Remember, the Ultra Deep Field is so small on the sky that 1,200 of these images would fit under your little fingertip. In any direction you care to face in the night sky, raise your smallest fingertip – you've just blocked out the light from over 18 million galaxies.

A truly gigantic stellar family

Most of the stars in our night sky are part of the family that make up our own galaxy, and while it is relatively straight-forward to chart the locations of the brightest stars, getting a more complete census of *exactly* how big our family of stars is, is a *much* more difficult task. Current estimates put the number of stars in the Milky Way at somewhere between 100 and 400 billion, which is not particularly precise. 300 billion stars is a pretty large number to be able to throw around, even by astronomical standards.

A factor of four, for a cosmic census, is doing par-ticularly badly. We can actually do considerably better for other, much more distant galaxies in our local universe. A factor of four is me telling you that your dog weighs somewhere between 60 and *240* pounds. A 60-pound dog may be medium to large, but a 240-pound dog is a differ-ent creature entirely. So why can't we do much better? It's because we're sitting within it. And because we're within it, we can't get a good viewing angle, to see what shape our galaxy really is, or where the stars are located.

So what can we do? A very basic path would be simply to try and count all the stars we can see, and see how far we can get. Some of the original maps of the galaxy were made this way, but by the time our estimates get all the way up to 100 billion stars, there's no way we're going to have enough time to count them. If we're right that there are at least 100 billion stars, and you only spend a second looking at each one, you're stuck with at least 3,180 years' worth of work laid out in front of you. That's not insurmountable, if you divide and delegate well enough, but there's another, more severe problem with the counting method.

The galaxy is full of gas and dust, both of which block out the light from more distant stars. Clouds of gas and dust tend to be the densest, and therefore the most opaque to starlight, within the disk of the Milky Way. Within the disk is smack dab where our solar system is, along with *most* of the stars in our galaxy. So not only are we most prone to missing stars as we look along the disk of the galaxy, but we're also missing the largest number of stars! Simply counting the visible stars would vastly underestimate the number of stars in our cosmic family.

So other than counting all the stars, what can we do?

There's a slightly more sophisticated method we can use to try to get around this 'dust blocks light' problem. In the optical regime, which is where our eyeballs function, dust will always obscure light coming from behind the dust cloud. If, however, we look in the infrared, the dust is transparent, and we're able to catch the glow of stars from both within the dust cloud, and beyond it. So if we can measure the total amount of light coming from our own galaxy in the optical, plus whatever it's doing in the infrared and other wavelengths, we can partially account for the light that might be missing. The trick with this method is that you now have to figure out how many stars should be responsible for the creation of a fixed amount of light.

To do this backwards calculation, you need to know how many of each size star you expect the galaxy to make. If your galaxy is good at forming very large stars, you need fewer stars to produce the same amount of light, because those large stars are also the brightest. If the galaxy is good at making very small stars, you could have a *lot* more individual stars, because the small stars are the faintest. You can add quite a few faint stars into the galaxy before you change the amount of light produced by a significant amount. Galaxies produce stars of all sizes, but the exact ratio of the number of small stars to the number of large stars is something astrophysicists are still trying to measure. We can make this measurement for the stars very near to us, but the fainter the star, the harder it is to spot. This uncertainty on how many large stars we should expect is partly why the total number of stars in our stellar census is so variable – and it's hopefully something we'll be able to nail down a bit better as research progresses.

But there's one member of the galactic family census that there's no uncertainty about, and we touched on its cousins in the last chapter: there is a supermassive black hole in the center of our galaxy.

Black holes in galaxies

As far as we have been able to determine, *all* galaxies contain a supermassive black hole at the very center of their bulge (see color plate 19). Unlike the black holes in Chapter 5, which form by the collapse of massive stars and are a few times larger in mass than our Sun, a supermassive black hole is a *billion* times more massive than the Sun. These black holes are a different class entirely.

All our observations suggest that black holes and their galaxies grow in lockstep with each other; smaller supermassive black holes are found in smaller galaxies, and larger supermassive black holes in larger galaxies. In addition, *one* supermassive black hole seems to be the rule. While it's possible to spot a galaxy with two gargantuan black holes at its core, this kind of scenario is almost exclusively encountered if the rest of the galaxy is looking particularly unusual. The assumption in these cases is that the double black hole galaxy has stolen its second black hole from an unlucky nearby galaxy that wandered too close and was devoured entirely. The unusual shape of the rest of the galaxy would then be the hallmark of the chaotic aftermath of having consumed that galaxy.

It's worth reiterating here that the black hole at the center of our galaxy isn't doing much of anything, other than sitting there and being massive. Much like our solar system is in a stable orbit around the Sun, the vast majority of a galaxy is in a stable orbit around its center, with no real reason to go plunging towards the black hole at its very center.

In fact, the region around a black hole can be quite barren. The Milky Way's central black hole, for instance, seems to be surrounded by stars but almost no gas, so there's nothing actively trying to fall into it. In order to be shredded by a black hole, a star would have to come extremely close to it. Some of the stars that orbit the black hole in the center of the Milky Way go around it once every *fifteen years* and we've been (incredibly) able to watch them move around it. Some stars come within a light day (16 billion miles, or 173 astronomical units) of the event horizon, and that's still not close enough to get torn apart or sucked in. (There are videos of the orbiting stars. Go watch them.* They are very cool.)

What's this about eating other galaxies?

Gravitationally driven collisions between galaxies aren't particularly common in our universe, but they're not particularly rare, either. Given that our universe is expanding, the relatively common nature of galactic collisions may seem surprising. This apparent contradiction stems from a set of simplifications commonly used to explain the expansion of the universe. We say, 'Imagine a balloon with a series of dots on the outside of it. Now inflate the balloon. All the dots move away from each other as the balloon, which is space, grows in size.'

Or perhaps, 'Imagine you have a loaf of bread with raisins in the surface. As the dough rises, the raisins will spread further apart from each other.'

....

* The actual data of the observations is hosted on the European Space Agency site here: http://www.eso.org/public/videos/eso0846j/ and a model view of that data is available here: http://www.eso.org/public/videos/eso0846h/

The fabric of space is indeed expanding as the universe ages, and these metaphors capture the concept of the distance between objects expanding, rather than the objects themselves moving at high speed. However, there's a tendency to illustrate this metaphor – the dots on the surface of the balloon, or raisins in bread dough – with regular patterns. We put everything on a grid, so that the effects of the universe's expansion are easier to spot. To some degree, this is convenient because it means we're only looking at one effect (the expansion), but it's a big oversimplification of the structure of the universe.

Objects in the real universe aren't laid out on a grid. The universe doesn't *do* grids. Real galaxies are scattered more irregularly across the fabric of space, which means that sometimes you're going to wind up with one or two or 50 galaxies pretty close to each other. Sometimes you'll wind up with a galaxy with nothing around it at all.

When you have two enormously massive objects relatively close in the universe, the force of gravity takes over. If two galaxies are near enough to feel the gravitational pull of each other, it doesn't matter that the universe is expanding – it isn't expanding fast enough to counteract the attractive force of gravity, and these two objects will inevitably fall towards each other. When they do, there's a good chance they will eventually become a single, larger galaxy, and the process gives us magnificent images (see color plate 21).

Interactions between galaxies can cause a galaxy to be strongly distorted, due to intense gravitational tides. Like the tides on Earth, these stretch out the galaxy, and pull material out into two 'arms'. Because these arms are caused by an external influence in the form of tidal forces, they are given the name 'tidal arms'. These tidal arms are often quite dramatic, stretching across many hundreds of thousands of light years. They are markers of a recent swing-by of the two galaxies, and often foreshadow a more dramatic collision of the gravitational companions in the relatively near future, where they will merge into a single, newly constructed object.

What of their black holes? After things settle down, the heaviest objects will end up in the center, which for two galaxies is their two supermassive black holes. Over time, the two black holes will lose enough energy while orbiting each other to merge into a single black hole, as we saw earlier. If the merging galaxy was about the same mass as the original galaxy, this should double the mass of the central black hole in one fell swoop – a much more efficient path to growing the size of a black hole than by trying to build mass with thin streamers of gas.

The future of our Milky Way

With all this talk of galaxies which are too close to each other eventually merging, and given how close we are to our neighboring, slightly larger galactic sibling of Andromeda (about 2.5 million light years away), we might well wonder what influence Andromeda has on our Milky Way's future.

The Milky Way is accompanied not just by the Andromeda galaxy, but by the entire Local Group, which contains some 50-odd galaxies. But Andromeda is by far the most significant of these, weighing in somewhere between 700 billion and a trillion solar masses. This is approximately the same mass as the Milky Way, which is also usually considered to have about a trillion solar masses' worth of stuff hanging around. The rest of the Local Group are mostly not very massive objects (like the Large and Small Magellanic Clouds), which are gravitationally tied to either the Milky Way or Andromeda, and which orbit the larger galaxy to which they're bound.

The Milky Way and Andromeda, as the largest objects in the Local Group, are also gravitationally tied to each other, and orbit each other extremely slowly. However, to determine where these orbits will take our galaxies in the future, you need to know a bit more about them than just their current positions and their mass. Both of those pieces of information are critical, but we also need to know how fast they're moving relative to each other. With this information, you can determine what path the two objects will take in the future.

The masses of two objects in space determine the point around which both objects will orbit. This is called the center of mass, and as we saw earlier, it is defined as the point in space that has an equal distribution of mass around it. For a system like the Sun and the Earth, the Sun contains almost all the mass – the mass of the Earth being so far away doesn't really change the center of mass very much. The Earth only pulls the Earth–Sun center of mass the tiniest bit away from the center of mass of the Sun itself. If the two objects are similar to each other in mass, however, the center of mass is between the two, in empty space.

This is the case for the Milky Way and Andromeda. They're both extremely massive objects, but neither can escape the gravitational pull of the other. They both orbit around a point somewhere near the middle of the space between them. This point – the center of mass between the two – is the location of an inevitable intergalactic collision in 3–4 billion years' time. Our two galaxies aren't moving fast enough relative to each other to avoid this point where our galaxy and Andromeda will eventually collide. Some time after that, the supermassive black hole which currently sits at the center of our galaxy, and the one which lives at the center of the Andromeda galaxy will merge together, to create an even more supermassive black hole.

It's tempting to think that our Sun and our solar system will be caught up and destroyed in this cosmic collision, but in truth the vast distances between the stars in both galaxies mean that no two stars will hit each other. The Sun's orbit around our newly merged, enlarged galaxy will certainly change, and the stars in the night skies of our planet will jumble, but the solar system will persist intact throughout the long process of the Milky Way's collision with Andromeda.

Quasars

The black holes at the center of the Milky Way and Andromeda are currently quiet, without any material actively falling towards them. But when a supermassive black hole does manage to gather a lot of material to itself, it can light up *very* dramatically.

Supermassive black holes can produce an absolutely stupendous amount of energy: *millions* of times the energy output of the Sun. This energy is a byproduct of the extraordinary inefficiency with which a black hole manages to absorb external material in order to grow. The material falling inwards gains a lot of energy, both in terms of the speed with which it is orbiting, and in heat. If only a tiny fraction of that material is actually absorbed by the black hole, the rest of it has to go somewhere. The energy produced by the black hole often funnels out into a bright jet of material, flung outwards along magnetic field lines at speeds that reach significant fractions of the speed of light.

These jets can only appear if the black hole is actively trying to accrete new material, and we might reasonably expect the galaxy it sits in to react somehow to the flinging of these jets. There's so much energy being produced that it seems only natural to think the galaxy should somehow care. This is still an active field of research, and we haven't quite converged to a consensus on how the black hole should shape the behavior of the rest of the galaxy. But a good place to look for such hints of interactions between the black hole's operations and the galaxy as a whole is in a class of galaxies with very extreme black hole behavior: quasars.

The term quasar is a portmanteau of 'quasi-stellar radio source', though nowadays the term has been retroactively turned into a portmanteau of 'quasi-stellar object', which then got abbreviated further into QSO.* The 'radio' part

---

★ Who says astronomers are bad at naming things?

got dropped out of the naming convention because it turned out to be a poor description.

Quasars are the observational signature of a particularly energetic black hole, doing a *very* good job of flinging material away from a galaxy. To an Earth-based observer, quasars are extremely bright point sources on the sky, looking very much like a star. Initially, these bright spots were puzzling, because a detailed investigation of their light spectrum revealed that they were like no star we'd ever seen. Because quasars blasted the camera with light, it was very difficult to determine their context – were they located in a specific kind of galaxy, or at a particular location in the galaxy?

It eventually became clear that these quasars were indeed in the centers of galaxies, and the galaxies they live within were incredibly faint relative to the brightness of the quasar, adding to the observational difficulty. The light from quasars had to be produced by the central black hole of these faint galaxies, which was dramatically outshining the rest of the galaxy. It also turned out that quasars are much more common at large distances from us than they are nearby – so they tend to exist much more commonly in the youngest galaxies in our universe.

How did they get so bright? To understand that, we need to poke our noses a little closer into the black hole.

An active black hole (in that it is trying and mostly failing to accrete new material and grow) is always surrounded by a superheated disk of material – an accretion disk – which is swirling around the black hole, gradually losing the angular momentum it needs to maintain a stable orbit, and slowly falling inwards. This innermost accretion disk is often surrounded by a larger donut-shaped ring of dust and gas, which is not quite as hot, and not nearly as close.

This disk of material will be surrounded by a magnetic field, like the Earth's magnetosphere or the Sun's, and so the easiest escape path for a particle is perpendicular to the disk, where the magnetic field can't get as tangled. Particles trying to escape find the easiest way out, which is always 'up' or 'down' from the disk, so they are funneled rather efficiently into a jet.

However, the particles that get fired out into this jet are mostly electrons and protons, not packets of light (photons). Getting from a beam of electrons to a beam of photons requires a few transfers of energy. One method of energy transfer happens when electrons travel along the untangled magnetic field lines extending outwards along the beam. The magnetic field causes the electrons to travel in a helix, as though they were tracing the path of an extremely long spiral staircase. These spiraling electrons give off high-energy particles of light; if the electrons are moving fast enough, they can spit out gamma rays. If they're moving less fast, they can radiate their energy away through X-rays and radio waves.

You can also create high-energy photons (up to gamma rays) with a more straightforward path: via collision. If you take an extremely speedy electron and crash it into a photon, the photon can gain enough energy to become a gamma wave. The end result of both of these processes is a pencil beam of high-energy radiation speeding outward in two jets away from the accretion disk of the supermassive black hole.

If you happened to be pointed in a direction that had you looking straight down the jet, you'd find yourself staring down a very bright portion of the black hole, instead of being able to see the glare from a less direct path. It seems that with quasars, we happen to be looking down the jet, or very nearly directly down it. The jets produced by quasars are very stable, and can reach lengths of hundreds of thousands of light years.

Orientation can't be the only key to spotting a quasar, because then we'd expect to see them nearby as well, and we don't. So there must be a secondary factor at work as well; very likely this is the amount of gas hanging around near the center of the galaxy. If there's not enough gas in the galaxy to create that innermost disk, the black hole simply won't have any material to fling outwards, much like the situation in the Milky Way. In the local universe, galaxies tend to be mostly stars, with less than 25% gas. But galaxies which existed at earlier times in the universe have the percentages reversed: they can easily be more than 60% gas, with fewer stars. This means that even without any external effects from things like collisions between galaxies, it's much easier for the black hole to have an easy fuel source to power its huge jets, streaming away from the galaxy.

Galactic gravitational lensing

There's one more trick that a galaxy can pull, and it works by nature of its mass. As we saw in Chapter 5, the larger an object, the more distorted the space around it, as space warps in response to the presence of mass.

If you try to pass a beam of light through the distorted space around a massive object, the distortion of space itself will cause light to curve its path. This is particularly noticeable around a black hole, but galaxies are also able to bend light. This bending is not so extreme as around a black hole, because the distortion of space surrounding an entire galaxy is less steep. A galaxy is not nearly as densely packed as a black hole.

Similar to the way that a magnifying lens changes the path of light and magnifies it, the gravitational distortion surrounding a galaxy can also function as a lens. Indeed, this bending of light is technically called gravitational lensing.

If you have a look at the objects along the edge of a photo taken with a fisheye lens, you'll notice that things that would be straight lines (like walls or trees) are bent into a curve – the light traveling through the lens was distorted as it passed through. The fun thing about a gravitational lens is that it means we can spot objects that would have been too distant or too faint to see without the magnification that the gravitational lens was able to provide.

The gravitational heavyweight that serves as the distorter, bending the light from the object behind it, could be a single, massive galaxy. (Equally, a cluster of galaxies, which collectively distort the space around them, can do the trick.) In order for us on Earth to see the lensing effect, the alignment of Earth, heavy galaxy, and distant object has to be exactly right. We want the light from the distant object to be almost exactly lined up, precisely behind the galaxy, from our viewpoint. The closer to a perfect alignment you can get, the more strongly distorted the light will be, as it tries to go through the most affected parts of the space surrounding the galaxy.

If everything is very *exactly* lined up, then the light from the background object will pass around the heavy object in front of it in an even way, like water flowing over a sphere. As we see it, there would be a perfect ring of light from the background object around the heavy front galaxy (or galaxy cluster). This is what's called an Einstein ring, because it's a perfect demonstration of the prediction of light's path made by Einstein (see color plate 23).

Most of the time, the background object, the heavy lensing object, and we as observers are very slightly out of line. This slight failure to align means that light isn't passing around the heavy lensing object in an even way. More light will bend around one side, leaving an empty space on the other side. If the background object is offset enough, you wind up with multiple distorted images from that object – they'll just look a bit stretched, like the fisheye lens effect. This is the effect that caused the 'Space Invader galaxy' (see color plate 22).

Usually, the background object behind the heavyweight is another older, more distant galaxy, and it's the light from this galaxy that is sheared out and stretched like silly putty. The foreground gravitational weight really just has to stretch space for this to work; but if you want a perfect Einstein ring, you have to cross your fingers and hope for a cosmic alignment.

✳

With galaxies strewn haphazardly around our universe, whether they be merging, forming stars, forming no stars, within a cluster of hundreds or thousands of other galaxies, or relatively isolated, by studying the way our galactic family is distributed, we learn about the next tier of our cosmic family tree – the fabric and structure of the universe itself. The universe as a whole is a surprisingly difficult piece of our family tree to learn more about, for all that we live within it. Nonetheless, our understandings of all phenomena are based on determining the rules governing the universe at large. To examine the universe, we must become time travelers.

7

THE UNIVERSE AT LARGE

The only home we will ever know

We've traced our way up from parent planet Earth, to grandparent Sun, to great-grandparent Milky Way, and now it's time to rise another step higher in our cosmic family tree to our cosmic great-great-grandparent: the structure of the universe itself. But what is left in the universe, which provided the means for galaxies to form?

The universe, as far back as we can measure it, is remarkably similar in all directions. It's not *perfectly* similar in all directions — otherwise we wouldn't expect to see individual galaxies, separated by wide distances. What 'remarkably similar' means is that the number of galaxies in any given direction seems about the same; and the galaxies which all have a particular shape aren't all over to the left and a different shape over to the right — the *types* of galaxies are also pretty evenly spread. So where did these galaxies come from?

In the very earliest light we can detect within our

universe, we observe a cloud of hot, dense, high-energy particles, currently measured at a fantastically even temperature of 2.725 Kelvin★ in every direction we look. This is reassuring, because if this cloud were very uneven, we'd expect that unevenness to be reflected in our current universe. However, if we look carefully enough, with precise enough instruments, we can measure very small temperature changes across the sky. These temperature wiggles are a reflection of tiny density fluctuations in the earliest universe. These tiny changes are small enough that purely random collections or sparsities of particles, found within this cloudy soup of high-energy particles, before the universe was cool enough to form atoms, was enough to create them. Once a density wiggle came into being, it collected more material to itself, growing larger, until that smallest of overly dense regions became the seed for the galaxies and all their stars.

....

★ 2.725 Kelvin is the equivalent of −270.4°C, or −454.7°F. Kelvin is a temperature scale like Celsius, but instead of having zero be the freezing point of water (0°C), zero Kelvin is an absolute minimum temperature. Absolute Zero is defined as zero Kelvin. 0°C corresponds to 273K.

Observing the universe

In trying to understand planets, stars, and galaxies, we've always taken the approach of looking at as many of them as we can, and then using the information about how frequently different styles of cosmic object have appeared to learn something about how they must have formed.

But we only have one universe to observe, and much like the problem of trying to figure out the size of our own galaxy, we're stuck inside that universe, and observing it is tricky, because our own position determines which part of the universe we can actually spot.

The biggest limiting factor is time. We can only record the portion of the universe surrounding us, where there has been enough time for light to leave an object, travel across space, and actually reach our human observers on planet Earth. As a cosmic speed limit goes, the speed of light is high, but considering the distances involved between galaxies, it starts to feel rather constraining. With a start date on the universe of 13.7 billion years ago, the most distant things visible to us have been sending light in our direction for roughly 13.7 billion years. But of course we also see those things as they *were* 13 billion years ago, not as they might appear if we could teleport to them *now*.

In the nearby universe, light's limitations are not a major constraint, as the time delays they impose are relatively manageable. The delays even to communicate with the New Horizons spacecraft, at the very outskirts of our solar system, are only a few hours (for light, it's a nine-hour round trip from Earth to Pluto and back). From Earth to Mars is a positively rapid 14-minute round trip on average. Within our galaxy, it starts to be a little more noticeable that our vision is delayed getting to us; for instance, our current observations of Eta Carinae indicate that that particular star is probably going to explode sometime soon. However, light takes 7,500 years for it to reach us, so if it had exploded last week, we'd still have to wait that long before we'd be able to notice.

The further out we go, the further our vision lags more and more dramatically behind 'now'. This lag affects light equally as it comes from all directions, so we end up with spheres of lagged space-time all around us. Because we are at the receiving end of all this light, this observable sphere is by definition centered on us. But, as we noted earlier, any other observer at any other position in the universe should see the same thing – their view of the universe will be just as limited in scope by the speed of light as our own.

The section of the universe that we can actually receive information from is known as the 'observable universe', and that's all we have to work with in deciphering how the whole thing works. This is not a catastrophic problem, because anywhere we look in our section of the universe, physics seems to work in the same way. Extrapolating outwards to say that physics should work in the same way even in the parts we can't observe is not an outrageous stretch of logic. Of course, we have no idea how *much* universe we're missing – or even if there's a number we could put to how much we're missing. Since we live when we do, not enough time has passed since the Big Bang for us to get information from more distant reaches of the universe.

The universe beyond our view may well be infinite. Sometimes saying that the space in our universe is infinite makes the mathematics of describing it a little easier to manage. However, we don't have an infinite amount of time to observe, in order to test the precise infinite or finite nature of the expanses of space, so we won't ever be able to determine the exact size of the entire universe.

Even with just our observable portion, though, we can predict a number of things about how the universe at large should behave, and test those predictions against other observations of our little bubble. It's through this method that we've untangled how the universe has evolved over time, which gives us information about how it began, and how it will unfold in the future.

We cannot observe the entire observable universe in one fell swoop, however, much as some astronomers (myself included) would love that. In practice, observations are undertaken through a number of paths, some more patchwork than others.

Each telescope facility gives a unique view out onto the universe, tailored to different scientific questions. Some may be optimized for hours-long exposures of small patches of sky, in order to capture the faintest traces of light from a distant planet or galaxy. These tiny patches of sky are chosen by individual scientists, and while most of the time there are checks in place to make sure that four people aren't repeating observations of the same area of sky, there's a *lot* of sky, and it would take a very long time to cover all of it.

The next strategy is to cover a slightly larger area of the sky, but not to sit on each part for quite as long. In general these fields are about the size of the full Moon on the sky, and many of them are named. The most famous of these is likely the Hubble Deep Field (and the smaller Ultra Deep Field), though they are both about ten times smaller than the full Moon (see color plates 20 and 24).

There's one more way to observe the universe – an all-sky survey. All-sky surveys try to look at the entire night sky. For telescopes on Earth, this is limited by where they're planted – a telescope in the northern hemisphere can't observe something in the southern hemisphere's sky. If you put your telescope in space, you can avoid this problem. Once you're there, depending on what wavelength of light your detectors are sensitive to, each telescope will return its own unique census of what's out there. These all-sky surveys range from Planck, which sent back the most detailed image of the universe's oldest light, to Gaia, which returned the very precise locations of the stars in the Milky Way. If you've seen an edge-on image of the Milky Way, that image was generated as part of an all-sky survey.

The universe is expanding

One of the most important discoveries that has come from studying the early universe and how it has evolved is the principle that the universe is expanding with time – and the longer we can wait around, the faster that expansion is proceeding.

It was not actually that long ago that scientists thought the cosmic family tree ended with our own galaxy – it was believed that all the objects in the night sky belonged to our own Milky Way. In the early years of the 20th century, this view began to change rapidly, and the family tree began to be populated with other galaxies. The work that was done in order to establish that our Milky Way did, in fact, have siblings, marked the beginnings of a new branch of astronomy – the study of other galaxies beyond our own.

Scientists began to take measurements of what we now know were other galaxies, trying to determine how far *exactly* they were away from us. However, a curious thing appeared in every measurement of these spiral and elliptical 'nebulae', as they were called then. Each one appeared to be moving away from us.

This particular measurement is a relatively straightforward one to undertake. The new stars which form within galaxies are usually surrounded by gas which didn't quite make it into the formation of the star. The newfound star can then heat the gas. This heated gas glows a very specific color, depending on what it's made of. Hydrogen, for instance, has a very bright, unique pink-red glow, and oxygen glows green in these environments. The colors are due to the exact wavelength of light that the atoms produce when an electron loses energy, and we know these wavelengths quite precisely from experiments here on Earth.

If the light reaching us from another galaxy is not exactly the color we would expect if we were measuring it on Earth, it means that there's been some movement between the object that produced the light, and us receiving it. If it's moving towards us, the expected colors will arrive bluer than they would have if there wasn't any motion. The peaks and troughs of the light wave are pushed closer together, making them a higher frequency, which registers them as a different color – in this case, blue. If the light wave is stretched out instead, the light will be shifted towards the red. Cunningly, we have termed these changes redshift and blueshift. It's exactly the same phenomenon as a doppler shift in sound waves, changing the tone of a siren as it approaches and then passes you.

So when astronomers looked out at the night sky, and found that *everything* was redshifted and not blueshifted, that implied that *everything* was moving away from us. And if everything was moving away from us, the universe couldn't be in balance. If the universe was in perfect balance, with no expansion or contraction, you wouldn't expect to see everything else fleeing from you as though you were at the top of a hill, watching everything roll away.

So either the universe wasn't in balance (in this case it would have to be expanding), or the Earth, our vantage point, was in a special place, like the top of our metaphorical hill. Historically, scientists haven't liked arguments that mean we have to exist in a statistically improbable location in the universe, and nowadays we have another argument at our sides that works against this.

No matter where we look in the sky, everything looks pretty much the same.

Sure, the small things change – there might be five galaxies in a group here, thirteen in a group there, but on average, galaxies are distributed nearly evenly across the sky. And, looking at the distribution of galaxies, it shouldn't matter where we're standing – if we were in a totally different galaxy, in a vastly different part of the universe, we should still see a universe that looks pretty much (at least on a statistical level) the same as ours. The technical terms for this are 'homogeneous' (the same from any point you could be standing) and 'isotropic' (the same in every direction you look).

A receding family of galaxies

The combination of the galaxies in our universe all appearing to recede from us, and our knowledge that we're not anywhere special in the universe, leads us to another question. If we're nowhere special, how is it that everything appears to be receding away from us? *How* is the universe expanding? Is there a 'center' somewhere that is feeding new space into our universe, or is there no center, and every galaxy is simply embedded in an evenly expanding universe? Fortunately, these two scenarios would reflect their changes both in the way we see galaxies moving away from us, and in the distribution of galaxies in the sky.

If the universe is expanding evenly, from everywhere, gradually extending the space between galaxies, then we would actually expect every galaxy to appear to be moving away from us. As a result, this is our current theory of how the universe's expansion is unfolding. If all things are separating, but we are stuck in a fixed location and not moving, then relative to us, all things are moving further away. Notably, this would appear to be true no matter which galaxy you happened to find yourself in – you'd always observe everything to be receding from you. If, on the other hand, space spooled out from some kind of vortex over to the left, we'd notice a difference in the way that galaxies were spread in the universe, and a difference in redshifts as we looked around at different parts of the sky.

Say we're looking directly at our spooling space vortex, and it's pushing new space out into existence. That wouldn't change the distance between our galaxy and the galaxy behind us, away from the vortex. So the galaxy behind us would appear to be stationary relative to us. In fact, most galaxies on 'our side' of the vortex would appear to be moving either slowly or not at all, relative to us. Things on either side of the vortex wouldn't appear to be drifting forwards or backwards, but they'd have some pretty solid apparent sideways motion – at high speeds, this might drive a measurable parallax,* where you could see its motion against the background galaxies. Meanwhile, galaxies on the other side of the vortex would appear to recede from us very rapidly!

I said we'd also notice changes in the distribution of galaxies – and that's because in this scenario, the vortex is forming empty space, so it's creating a perfect bubble: an absolutely empty sphere of space, growing rapidly with time. And if we were to take large-sky surveys of the galaxy population, such a bubble would surely stand out as unusual. We've actually looked for signs that we might be inside such a cosmic bubble (though, clearly, it's not as empty as the vortex-bubble would be), and so far we've found not even a hint of evidence that we're in one.

Instead, data like galaxy counts, temperature evolution over time, and density measurements out to very large

---

* Parallax is an effect by which nearby objects will appear to move against distant objects in the background, which appear to remain stationary. Any two lines of sight can produce this effect. Holding a finger out at arm's length with one eye closed, and then closing that eye and opening the other, you will see your finger appear to jump sideways against the background beyond it. In astronomy we typically use the position of the Earth six months apart (so that the Earth has moved by twice the distance between the Earth and the Sun) to spot the motions of nearby stars against the more distant stars which remain unmoving.

distances point towards the expanding, isotropic, homoge-neous universe picture, and not to the Earth being at the privileged center of a vortex (or explosion). If there *were* a cosmic vortex somewhere, spooling out space-time from a specific point, ultimately we would expect the way we observe the universe to be very different. Instead, because space is being created between all the objects that already exist, we observe all galaxies to be moving away from us. We also observe that the universe is distributed evenly across the sky, so no matter where we stand, we should see something similar.

The distant universe is a smaller place

So if the universe is expanding the longer we wait, if we rewind time, we should find ourselves in a smaller, more compact universe. This is indeed the case, and if we com-bine this feature with the limitations of the speed of light, it means that the most distant light we're able to observe is also coming from a period of time when the universe was significantly more dense than it is now.

If we think about the observable universe as a series of shells surrounding our Earth, where each shell gives us a different window in time on our changing universe, then another feature of this time-traveling view is that the volume of space we can observe with each shell increases the further back in time we go.

Light's shift towards the red, used to measure whether objects are moving away from us, and if so, at what speed, is useful here. Intrinsically, the redshift of an object tells you how far the light has been shifted from its rest position. But since the most distant galaxies are also the ones that have the largest shifts to their light, redshift can also be used as a metric for the age of the universe. Since we've figured out how fast the universe expanded, and when that expansion happened, we can convert the age of the universe (from redshift) into a rough size of the universe when the light left a distant object. At a redshift of 2,[*] the universe is a little less than 3.5 billion years old,[†] and one third its current size. A galaxy at a redshift of 9 is living in a universe of one tenth its current size. Add one to the redshift, and then make that into your fraction. (This math is a bit of a rough estimate, but it's a good way to get a general perspective.)

....

[*] Redshift is a ratio of the shift of the light as observed, relative to the expected wavelength of the light if the object weren't moving. For instance, if the wavelength of light in a lab was 656 nanometers, but it had been shifted to 1,968 nanometers because of the expansion of the universe, the difference from the expected is 1,272 nanometers. Dividing out by 656 gives us a redshift of 2.0.

[†] This means, in turn, that the light from objects at this distance has been traveling for a little over 10.5 billion years.

Since we are looking back to a universe that is physically smaller, this means that the distances between galaxies are all smaller, and the entire universe at that earlier time must be more dense than it is now. But the critical thing to consider here is the *volume* of space we're able to observe. Things that are very near us we can only see within a very small volume; at greater distances we see a much larger volume of space. This means that as we look back at more distant objects, we are seeing a larger *fraction* of the universe, the further back in time we look. Since we don't know the total volume of the universe, it's impossible to say how much that fraction changes, but it's certainly a bigger number than for the nearby universe!

Our Milky Way's physically closest galactic family members fill only a very small fraction of the current universe. If we try to push to galaxies beyond this immediate neighborhood, we recede backwards in time, but we expect that the volume of space that operates close to our 'now' should be repeated many, many times over to fill the universe that we cannot observe but know must be there. Another galaxy's family neighborhood may be filled with more galaxies, or fewer companions, or no companions at all.

This shifting area that we can survey in the universe becomes convenient for the sorts of galactic family censuses that scientists might like to undertake, to try to count how frequently certain types of galaxies or configurations of galaxies appear. These studies of the galaxy population are much easier with large numbers of galaxies, and it's easiest to catch large numbers of galaxies when you have a large volume to look into. The very local universe doesn't afford us the luxuries of large volumes, but at the furthest distances, we can start to say how common groups of galaxies should be.

It's much harder to say whether or not it's uncommon for a galaxy as large as our Milky Way to have as large a companion as Andromeda at this age in the universe – our observable universe only has a few galaxies which are close enough to be in the very recent past as we look at them. The shell of space which surrounds us, before the time lag from light's travel time begins to be large, is reasonably small. We can't go hunting for other Andromedas around other Milky Ways in the universe 'now', because we begin to go too far back in time the more distantly we search.

Not all galaxies are separating

We must remember, of course, that the galaxies within the universe are not evenly spaced out. Ours is a universe that looks rather spidery, with filaments of the cosmic web stretching across the sky. There are grand areas of nothing, and clusters of thousands of galaxies, connected to other clusters with faint tendrils, each made up of glowing galaxies.

When we say the universe is expanding, we do in fact mean that the space between objects is increasing, but the web-like distribution of galaxies means that this is not quite as simple as extending the distance between *all* objects in the universe.

There are two reasons why this is so, and the first is that the force driving the universe's expansion appears to be very, very small. There's an awful lot of space in the universe as a whole, so as a bulk property of the universe, we measure a significant change, but in smaller regions of space the effect isn't as strong. The second reason is that in those smaller spaces, the force of gravity is much, *much* stronger than the force pushing objects apart. Groups of galaxies, galaxies themselves, and certainly solar systems and planets, are not being pulled to larger sizes by the expansion of the universe at large because the force of gravity holds them much more tightly together.

Gravity is a bit of a juggernaut in the astrophysical world, and if two objects are gravitationally bound to each other (meaning that their relative speeds are too slow to let each escape the pull of the gravity of the other), cosmic expansion isn't going to be able to do much about it.

The expansion of the universe would have to be frighteningly enormous (much larger than we observe it to be) to pull apart two galaxies which are bound to each other by gravity.

In the last chapter, we talked about the future collision of the Milky Way and Andromeda. We can say with such certainty that they will collide because they are gravitationally tied to each other – the expansion of the universe will not draw the two of them apart. Any bound objects in the universe won't be affected by this expansion, so not only are the Milky Way and Andromeda immune to their universal parent's continual expansion, but the entire cluster of galaxies to which our galaxy belongs will remain in place. However, galaxies in a neighborhood of the universe which are not gravitationally tied to our galaxy or its cluster of companions will find themselves drifting ever further from us over time.

Our cosmic lineage

To make a usable map of our cosmic lineage is quite a tricky thing. From our vantage point on Earth, we have mapped out a list of relationships between our planet and the others within our solar system, between the Sun and the other stars, and between our Milky Way and the other galaxies. With every tier of our family tree, we have tied our understanding to our own home, whether that be planet, star, or galaxy. But to go forward from a description of family ties to a detailed mapping of where we actually stand requires not just information on respective positions, but measurements of the distances to each of our family members as well.

We know our cosmic lineage quite well up to a point. You, reading this book, have a street address, in a country, on planet Earth. Earth is part of our solar system of planets, which orbits the Sun. The Sun is one of many billions of stars which orbit the center of the Milky Way galaxy. The Milky Way, in turn, orbits the center of mass of the Local Group, somewhere in between the Milky Way and Andromeda. The Local Group sits within the Local Supercluster, a loose association of other clusters and groups of galaxies, spanning hundreds of millions of light years.

But this kind of lineage is only useful if you know where to find the Supercluster. If you know where any individual part of our lineage is located, you might be able to trace your way back to find our planet from elsewhere in the universe, but without that anchor to a different set of knowledge, it's not particularly helpful or informative.

Scientists have grappled with this problem on a much smaller scale before. Both the Pioneer and Voyager spacecrafts tried to relay our position in the unlikely event that they were ever encountered and picked up by another intelligent race of beings. The Pioneer plaque showed the location of our planet within our solar system, but it (and the Voyager golden record) also provided a map that would be useful to someone else from within our galaxy. The set of lines radiating out from a single point on the left-hand side of the Pioneer plaque and the Voyager record is this map. This gives the positions and frequencies of a set of pulsars – incredibly rapidly spinning objects that beam light out into the cosmos like a lighthouse. Each pulsar has its own unique frequency: the number of times per second it flashes in our direction. Pointing out the distances and frequencies of fourteen pulsars, as we did, should allow someone else, observing the same set of pulsars, to triangulate our position.

However, this can only be scaled up so far. If you're trying to tell someone outside our galaxy where we are – or really, anyone sufficiently far away in our universe – you swiftly become tangled in a different problem, which is that the further from our planet you would like to map out for our faraway friends, the more you travel back in time. At some 'distance' from our planet, our exploration becomes one much more of time than of space. Even just the Virgo supercluster (a tiny corner of the universe) is some 110 million light years across. Between light leaving that side of the cluster and our receiving it, things have changed over there in the intervening 100 million years. 100 million years on an astronomical timescale is not a huge amount of time, but it's much longer than humans have inhabited the Earth.

This time traveling becomes an increasingly disruptive problem the further back in time we go. At some point, we are looking at structures and objects that no longer exist. Many things can happen in a couple of billion years, and they usually do, so using those objects on our map isn't very useful for an intergalactic or universe-wide traveler.

Unfortunately, there's no solution to mapping out our lineage in a foolproof manner so that anyone, anywhere, would be able to point to the planet where the humans reside. One of the fundamental tenets of cosmology tells us that there are no 'special' perspectives on the universe. This means that there's no overall frame of reference that every possible observer could agree on, no universe-wide 'north' by which to orient our maps.

The best thing we may be able to do right now is to create a really good map of our little local set of galaxies and clusters, and tell any visitors, if they happen to see something that looks like it in their travels, that we humans are the children of this particular part of the universe.

INDEX